Agrarian Structure and Tenancy Movements

Agrarian Structure and Tenancy Movements

G.V. Joshi

Rawat Publications
Jaipur and New Delhi

ISBN 81-7033-580-9

Published by
Prem Rawat for ***Rawat Publications***
Satyam Apts., Sector 3, Jain Temple Road, Jawahar Nagar, Jaipur - 4 (India)
Phone: +91 141 651748 / 657006 Fax: +91 141 651748
e-mail : rawatbooks@pinkline.net

Delhi Office
G-4, 4832/24, Govind Lane, Ansari Road, Daryaganj, New Delhi 110 002
Phone: 011-3263290

Typeset by Rawat Computers, Jaipur
Printed at Nice Printing Press, New Delhi

Dedicated

to

Dr. Dinakar Desai (1908-1982)

who pioneered the Tenancy Movement

in

Uttara Kannada District

Contents

Foreword 9
Preface 11
Glossary 13

1 Introduction 17
2 Early Agrarian Structure and Factors Behind Tenancy in Uttara Kannada 61
3 Tenancy Movement and Legislations in the Pre-Reorganisation Period 95
4 Tenancy Movement and Legislations in the Post-Reorganisation Period 133
5 Summary and Conclusions 161

Bibliography 175
Index 185

Foreword

Tenancy movements constitute a problem for study in social sciences as a whole, and not in economics, political science or sociology alone. Approaching this from the narrow angle of a single social science discipline would not have been satisfactory. It is to the credit of Joshi that he has ably incorporated economic, political, social and historical aspects of tenancy movements in Uttara Kannada in this book *Agrarian Structure and Tenancy Movements*. It is this interdisciplinary perspective which makes this work a valuable contribution to the literature on tenancy movements and reforms.

Major issues like the factors behind agricultural tenancy, the forms and terms of tenancy and the regional variations in the extent of tenancy are analysed by Joshi within a theoretical framework. He has been able to disprove some of the premises held by the scholars who are generally taken as the proponents of the neo-classical theory. Joshi brings out the fact that to comprehend all the historical, economic and social factors that shape the dynamics of agrarian structure and also to see to what extent the goal of equity is accomplished through tenancy movements and reforms, we have to rely mainly on the political economy approach.

Joshi's study shows that the secret of successful implementation of tenancy reforms in Uttara Kannada district does not lie in the enlightened magnanimity of bureaucracy and political leadership ruling from the capital, but in the mobilised peasantry itself who fought for their rights. The tenancy movement in the district, exposed effectively the contradictions between the landlords and the tenants, gave expression to the urgent need for tenancy reform and also created necessary climate for its implementation.

Joshi has taken immense pains to marshal historical documents from various sources, and also to supplement this historical analysis by personally interviewing leaders who had been active in the tenancy movements so as to have a complete picture. Had he relied only on desk research, he would not have come out with so many insights. The outcome, therefore, is an authentic document on tenancy movements and reforms in Uttara Kannada, which is bound to remain a source of reference for years to come.

Joshi's style is narrative, and his analysis does not rely on mathematical or statistical models. The simplicity of this presentation and the details including individual cases he likes to quote in support of his arguments, constitute, in my view, a merit of his work, and not a weakness.

I wish his book an excellent reception.

M.V. Nadkarni
Vice-Chancellor
Gulbarga University

Preface

This book is a part of my Ph.D. thesis "Agrarian Structure and Tenancy Reform: A Case Study of Uttara Kannada" written under the supervision of a distinguished economist Prof. M.V. Nadkarni, who is at present the Vice-Chancellor of the Gulbarga University. I am happy to state that Prof. Nadkarni is an illustrious son of Uttara Kannada itself. This work would not have taken the present shape but for the inspiring guidance and tenacious supervision of Prof. Nadkarni at the Institute for Social and Economic Change, Bangalore. His deep understanding of the tenancy problem in the region under study made our discussions extremely rewarding. Mere words would not be sufficient to record my deep sense of gratitude to Prof. Nadkarni.

Prof. D.M. Nanjudappa, Deputy Chairman, Karnataka State Planning Board and Chairman, Indian Council of Social Science Research was my esteemed teacher in the P.G. Department of Economics, Karnataka University, Dharwad.

He has always been extending academic support to me. I remain grateful to him.

The late Dr. Dinakar Desai, who pioneered the tenancy movements in Uttara Kannada, was kind enough to spare his time to provide necessary information even though he was not having good health. It is my duty to remember his encouragement. Both Shri S.V. Pikle and Shri Dayananda Nadkarni who were involved in tenancy movements in Uttara Kannada also provided necessary information. I thank them.

I was born in Uttara Kannada. Having had my early and college education in this district, I worked as Lecturer in Economics at Divekar College of Commerce at Karwar, the headquarters of Uttara Kannada, for a period of 11 years. I took to journalism (1974-79) in this district which, besides enabling me to develop wide social contact, helped me to understand and appreciate the problems of the district. While sending news reports about the implementation of the Karnataka Land Reforms Act of 1974, I developed an interest in the tenancy problem. The people of Uttara Kannada have been constantly encouraging me in various ways. If the present book turns out to be a modest effort at least in repaying a small portion of the huge debt that I owe to my district, I will be indeed happy.

I express my deep sense of gratitude to the Institute for Social and Economic Change where I could get many books, journals and reports throwing light on agrarian structure, peasant movements and land reforms.

I shall be failing in my duty if I do not thank Rawat Publications, Jaipur, for their kind consent to publish this book. With the publication of this book, my long-cherished dream has been fulfilled.

G.V. Joshi

Glossary

Inam Tenure The intermediary tenure characterised by almost absolute control of the intermediaries on the lands, whether cultivated or otherwise, in their estates. The Inam landholders paid stipulated revenue to the Government and exacted rents as well as a variety of cesses and levies, legal or otherwise, from the under-right holders. They had under them many under-right holders, with varying rights in the land, and not all of these were actual cultivators. See the details in *The Report of the Committee Appointed for Evaluation of Land Reforms in Maharashtra*, Government of Maharashtra, Revenue and Forest Department, Government Central Press, Bombay, 1974, p.234.

Mulvargadars Original proprietors of land in Uttara Kannada district.

Mulgeni	Permanent tenancy with hereditary rights found mainly in the case of garden land. *Mulgeni* tenant was popularly known as *Mulgenidar*. He was a tenant under *Mulvargadar*.
Nadagi	A type of quasi-permanent tenancy existing on a few coconut plantations in the coastal villages of Kumta and Honavar talukas of the district. The landlord had to bear the expense of planting the trees, and the *Nadagi* tenant, popularly called *Nadagidar*, had to bear the expenses of rearing them. The rent paid by the *Nadagidar* was generally fixed at one-half of the produce, and in a few cases, in cash.
Sulgi	Another type of quasi-permanent tenancy found on a few coconut plantations in Kumta and Honavar talukas of the district. When full-grown trees were made to the tenant, popularly called *Sulgidar*, the tenancy was called *Sulgi*. The *Sulgidar* used to get only one-third of the produce of land.
Chalgeni	Tenancy-at-will. It was the predominant form of lease in the case of paddy land in Uttara Kannada.
Palu	Share-cropping tenancy found in the case of paddy land in Uttara Kannada.
Korpalu	The system of share-cropping found in some villages of Haliyal taluka of the district.
Gazni	Paddy field located on the mouth of rivers and small creeks. In the absence of proper arrangements, it is susceptible to flood from sea-water particularly at the time of high tides.
Makki	Upland paddy field. It is a poor quality land.

Abbreviations Used

ATRS	Ankola Taluka Ryot Sangh
BTAL Act	Bombay Tenancy and Agricultural Lands Act
EPW	Economic and Political Weekly
KDKS	Kanara District Kisan Sangh
KDCRS	Kanara District Congress Ryot Sangh
KWT	Kanara Welfare Trust
NKDKP	North Kanara District Kisan Panchayat
NKLA	North Kanara Landlords' Association
OFESML	Owner Former Establishment Special Measures Law
PSP	Praja Socialist Party
RK	Ryot Koota, i.e., Peasants' Association

1

Introduction

Statement of the Problem

The Karnataka Land Reforms Act of 1974 was regarded by some as one of the most radical land reform legislations in India, more advanced than even the one in West Bengal at least in important respects.[1] The most important provision in the Act was the abolition of tenancy. Assumptions behind tenancy abolition were: (i) tenancy does not provide incentive for investment and production in agriculture, and (ii) the best way of ending the exploitation of tenants was abolition of tenancy, by converting tenants into owners. In a few other states like West Bengal where also radical reforms were implemented, the emphasis was on securing the rights of tenants rather than on abolition of tenancy. Often, therefore, the Karnataka legislation of 1974 was said to be more radical.

The nature and extent of tenancy is not the same everywhere. In Karnataka, in particular, tenancy was a dominant feature only in three districts, namely, Uttara Kannada, Dakshina Kannada and Shimoga. Though the assumption that the tenant is always a weak partner and that landlords belong to a dominant class need not always be true,[2] at least in the two coastal districts, viz., Uttara Kannada and Dakshina Kannada and the Malnad district of Shimoga, tenancy conformed to the traditional pattern assumed under tenancy reform. Table 1.1 reveals that in these districts there were not merely high levels of tenancy but also that this was more prominent in small holdings, particularly in Uttara Kannada.

Table 1.1

Extent of Tenancy in Selected Districts of Karnataka, 1970-71

(Size-Group-Wise)

Holding Size-Group (in Acres)	*Uttara Kannada*	*Dakshina Kannada*	*Shimoga*	*All Karnataka*
(a) Proportion of Tenant Holdings in the Total Number of Holdings in the Size-Group (%)				
Less than 5	70.4	55.4	21.1	10.3
5 to 12.5	60.8	55.0	23.3	10.4
Above 12.5	60.7	34.6	21.1	12.9
All Holdings	68.2	54.4	21.5	11.2
(b) Proportion of Leased-in Area to Total Area in the Size-Group (%)				
Less than 5	58.5	53.8	18.7	8.6
5 to 12.5	41.4	47.1	15.1	7.0
Above 12.5	30.7	21.0	7.2	7.2
All Holdings	43.4	43.6	13.4	7.4
(c) Proportion of Total Area (Leased-in and Owned) under Tenant Holdings to Total Area in the Size-Group (%)				
Less than 5	71.3	57.8	21.6	11.1
5 to 12.5	60.4	53.9	23.2	10.3
Above 12.5	61.0	31.6	18.8	13.8
All Holdings	64.0	50.2	21.2	12.3

Source: Derived from the Census of Agricultural Holdings in Karnataka, 1970-71.

In this sense, the crucial relevance of tenancy reform in Karnataka can be said to have been confined mainly to the two coastal districts and Shimoga (see Map 1). Elsewhere, it had marginal significance if any, particularly since it was known even before the Karnataka Land Reforms Act of 1974 came into effect that implementation of ceilings was not going to be a success. Subsequent studies confirmed this. Pani showed that unlike the high rates of implementation in granting land to the tenants under the provisions relating to tenancy abolition, very little land was redistributed under the provisions pertaining to ceilings according to government's own data. In as many as six districts, not a single acre of land was redistributed until October 1980. In thirteen of the nineteen districts in the state, the area actually redistributed was less than 20 per cent of the surplus land estimated by the State Government.[3] The analysis of land reforms statistics by Rajan also showed that the ceiling law touched a very small number of landowners, roughly 9550, despite the large scale administrative endeavour. The entire area of the surplus land, 2.73 lakh hectares, secured for the redistribution programme came from these landowners, three-fourths of whom were in just four districts of the northern Maidan region (Raichur, Bijapur, Gulbarga and Dharwad).[4]

If, therefore, land reforms (tenancy reforms) had a chance of success, it could be mainly in respect of tenancy in the three districts referred. As such, these districts particularly Uttara Kannada, offered a scope for evaluating the success and adequacy of tenancy reform because this district witnessed a fairly long and active peasant movement against tenancy and its oppressive character. Besides, a study of agrarian structure and tenancy reform at the regional level has its own justification. In the course of his review of land reform studies in India, Joshi had suggested: "...in the earlier period, the analysis of the agrarian problem has generally been done either at the all-India

Map 1
Karnataka State: District Boundaries

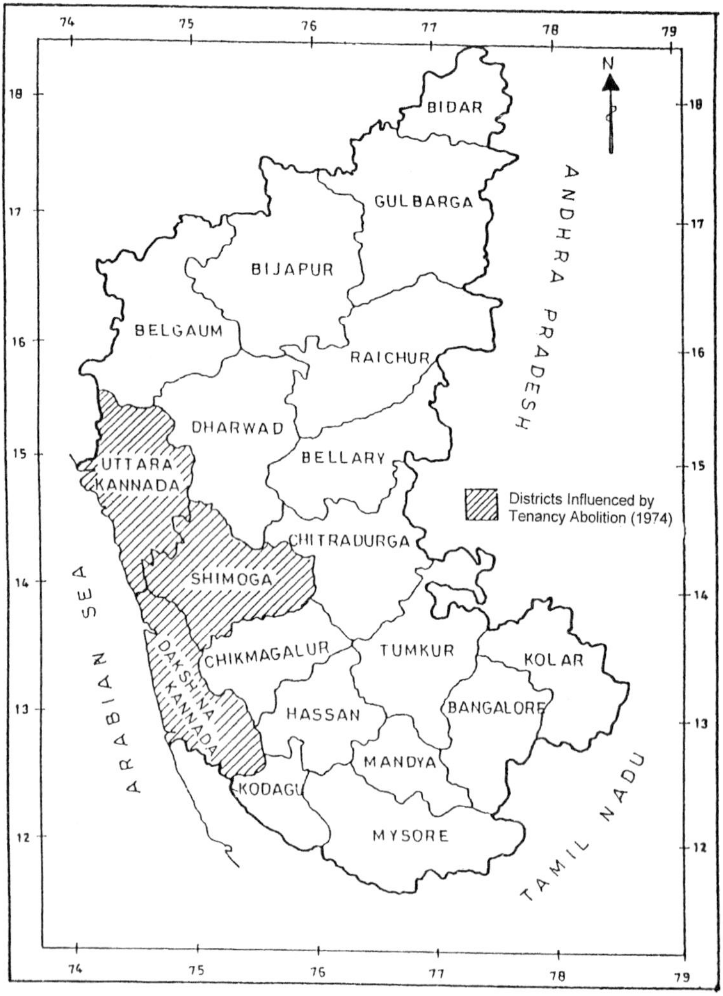

or state levels. It is now time that we make a shift to analysing the problem of land reforms at the level of regions within a state and more specifically at the level of groups of districts or even sometimes at the level of a single district."[5] This is because, at the level of aggregation of the state, regional features of agrarian structure get averaged out and blunted, and the micro-level problems of land reform can hardly be studied. This study attempts to analyse the agrarian structure and the role of peasant movements (tenancy movements) and reforms in changing this structure in Uttara Kannada, a single district.

Conceptual Clarifications

Before setting forth the theoretical context of the study, we shall attempt to define and clarify the main concepts used by us, namely, agrarian structure, tenancy, tenancy reform, agrarian capitalism and agrarian classes.

Thorner[6] and Joshi[7] use the term 'agrarian structure' to refer to the network of relations among various groups of people connected with or involved in the process of cultivation, viz., landlords, tenants and agricultural labourers. Beteille writes that "the study of agrarian system will centre around the problems of land and its utilisation for productive purposes."[8] He also recognises the interrelatedness of the structural elements in agriculture such as technology, work cycle, organisation of production and agrarian hierarchy. For Todaro,[9] agrarian system is the pattern of land distribution, ownership and management and also the social and institutional structure of the agrarian economy. Thus, while Thorner and Joshi use the term agrarian structure in a restricted sense, Beteille and Todaro hold a wider conception.

We prefer to use the term 'agrarian structure' in a wider sense rather than restricting it to merely the mutual relations of landlords, tenants and agricultural labourers. Besides, we emphasise the need for studying the dynamics of agrarian structure in a historical perspective. Cohen brings out, in an unequivocal manner, the role of historical factors in the formation of agrarian structure. To quote: "An understanding of the historical process behind the present state of a particular country is a first requirement for a realistic estimation of the agrarian structure there and a realistic evaluation of the possibilities of agrarian reform."[10] Following Todaro's conception and changing it to suit our purpose, we use the term 'agrarian structure' to refer to the pattern of land distribution, ownership and management and also institutional structure of the agrarian economy which is shaped by historical factors.

Tenancy generally implies relationship between the landlord and the tenant. Some studies like that of Bardhan which focus greater attention on sharecropping tenancy in India have brought to light its changing character, with increasing prevalence of cost-sharing by the landowner, particularly in the purchase of new inputs.[11] Bardhan also states that the dividing line between a crop-sharing tenant and attached labourer has been rather thin all along in any case, particularly where the landlord has been taking all important decisions on crop-sharer's farm. The important difference, however, is that the former, unlike the latter, holds some land on which he exercises some freedom of decision-making. Gradually, even this freedom may vanish and he is reduced to the status of a labourer, and his crop-share may be no more than a payment of wage in kind. In a dynamic setting, the sharp conceptual difference made between a tenant and a labourer disappears in course of time, even if they were valid in a classical or traditional setting in the past. Even, in the traditional setting, as Hanumantha Rao has contended, rents collected by the landlord were more or less equated to the surplus that they would alternately get by cultivating the lands themselves through hired labour.[12] Broadly, therefore, the landlord hired in labour by leasing out land, and the tenant hired out labour by leasing in land. Yet, the status of a tenant has been superior to that of an attached or bonded labourer, since tenants were not complete slaves to landlords in all respects and had relatively more freedom in decisions and also some right to share the fruit of these decisions.

The 'traditional and accepted' notion of land reform is to view it as a means to redistribute property rights in land for the benefit of the small holders and agricultural labourers.[13] Dandekar and Rath argue, however, that the essential purpose of tenancy reform is expansion of owner-cultivation and not redistribution of ownership or the cultivated holdings.[14] They admit that when legislations for realising the "land-to-the-tiller" ideal are enacted and implemented, the ownership holdings are redistributed but this redistribution is only a step towards expanding and promoting ownership cultivation. They also say that tenancy reform is a measure to secure for the tenant a larger share in the product of land he cultivates and to provide greater incentive for land improvement and greater production. Thus, the justification of tenancy reform traditionally has been on the twin grounds of equity/distributive justice and efficiency/productivity. Therefore, we conceptualise tenancy reform as a legislative measure which, by strengthening the legal and technical position of the tenant and enabling him to secure a larger share in the product of the land, aims at the objective of providing him greater incentive for investment and production in agriculture. It could be either through a tenancy regulation or its abolition.

The question of development of capitalism in Indian agriculture has given rise to a serious debate among scholars with which we are not directly concerned.[15] Notwithstanding the apparently conflicting views expressed by them, it is possible to identify some degree of consensus. Though Utsa Patnaik emphasised the capital intensification, nowhere does she say that the employment of wage labour and commodity production are not associated with the development of capitalism in agriculture.[16] Both Thorner[17] and Rudra[18] admit that the application of hired labour on a large scale is a characteristic feature of agrarian capitalism. The ability to raise agricultural yield substantially, taken by Thorner as a distinct feature of the capitalist farmer, obviously results in a large marketable surplus which Rudra also highlights. Thus, we can take employment of labour on a large scale, production of marketable surplus and its realisation through sale as operationally meaningful criteria for identifying agrarian capitalism.

As regards the criteria for defining agrarian classes, we follow Bardhan's scheme[19] and modify it to incorporate tenancy to suit our purpose. We propose here to identify four agrarian classes, namely, the landlords, rich peasants, poor peasants and the landless agricultural labourers. The landlord is a person who gets his land cultivated either through hired labour or by a tenant. In other words, he always hires in other's labour power and never puts in self-employment. Symbolically, in the case of landlord, SE=O, HI>O, while HO=O (SE= self-employment, HI=hired-in labour, HO=hired-out labour). The rich peasant, as a rule, relies on hired labour or tenant's labour, apart from putting his own labour on land. He may hire out his own labour as a tenant, but never as an agricultural labourer. Thus, in his case, SEO, while HI>HO. The poor peasant holds land, but survives mainly by hiring out his labour. He may be either an owner-cum-labourer, or an owner-cum-tenant-cum-labourer or a tenant-cum-labourer. He may hire in labour to carry out seasonal agricultural operations. While in the case of the rich peasant HI>HO, in the case of the poor peasant HO>HI. The landless labourer does not hold any land even in the capacity of a tenant. He makes his living only by hiring out his labour as a labourer. In his case HI=O, while HO > O. It is quite possible that the landlords and rich peasants, unlike the poor peasants and the landless labourers, may earn additional income by undertaking non-farm occupations. They may also gain by indulging in usurious practices.[20]

Some scholars who emphasise the sociological perspective of the Indian agrarian structure hold the view that the caste factor plays a vital role in shaping its dynamics.[21] Finding a close interrelation between the Hindu caste hierarchy and agrarian structure, they state that the landlords belong to the so-called upper castes, the tenants to the middle castes and the landless labourers to lowest castes. This situation, they say, leads to economic disabilities being augmented by social disadvantages. Beteille, an eminent sociologist himself, does not accept this view; he argues that it is the concept of class, and not caste, that provides proper analytical tool to study the system of agrarian relations in India. To quote him: "It may appear ironical that while the architect of the concept of class believed it to be particularly suited to the study of industrial societies, the experience of sociological research may well suggest that it is in fact more useful in the study of agrarian structure than of industrial communities."[22] He further tries to show that the caste which has control over the use of land is the dominant class in rural India irrespective of its ritual status. Although we agree with him in principle, we still argue that the caste-class debate ceases to be relevant in a region where both caste and class coincide.

Theoretical Context

We can identify two different theoretical approaches followed by the different scholars in studying tenancy and agrarian structure: (1) neo-classical or liberal approach, and (2) political economy approach. The neo-classical approach is based on tacit assumptions like basic rationality or self-interest of the peasant and freedom of choice or economic decision-making, in the light of which prevalence of tenancy, its forms and terms are tried to be explained. The political economy approach, on the other hand, emphasises the role of institutional factors, particularly the class composition and dominance in shaping the agrarian structure and development of a region in a historical perspective at the stage of implementation of reforms. The significance of the present study lies in the fact that it is based on the political economy approach, the perspective of which is much wider than that of the neo-classical approach.

Adams and Rask[23] and also Khusro,[24] who may be taken as the followers of neo-classical approach, attribute high level of tenancy in the less developed countries to increasing pressure of population on land and the relatively inelastic supply of arable land. But the instance of Japanese agriculture clearly indicates that tenancy cannot be simply attributed to growing pressure of population on land alone. In the pre-reform period, in Japan, the total tenanted area constituted as much as 46 per cent of the total cultivated area even though farm population had remained absolutely stable for a long period of 60 to 70 years.[25] The high level of tenancy in Japan was indeed the result of institutional factors like the glaring inequality in the ownership of land and urban based landlordism.[26] The tenancy of capitalist nature, of course, was widely prevalent even in the United States in the 1960s.[27] Thus, clear is the evidence to show that tenancy is not the consequence of only demographic pressure on land.

Hanumantha Rao, while admitting the significance of demographic pressure, stresses the role of natural factors like fertility of the soil and assured water supply, either in the form of rainfall or irrigation in giving rise to tenancy relations.[28] He argues that land should be fertile enough to yield sizeable surplus to support the tenant as well as the landlord. To put it differently, the two parties enter into tenancy contract because it is profitable to both.

We argue, however, that the natural factors themselves cannot adequately account for a large number of tenancy contracts in any region. Given the importance of natural advantages, the level of tenancy is likely to be very high in a region where the crop grown is labour-intensive and agricultural operations are tiresome and arduous. Considering this reality, Beteille points out that aversion to manual work is stronger in areas of wet paddy cultivation than in the areas of wheat cultivation. Citing the instances of Tamil Nadu, West Bengal and Kerala, he states that the aversion to manual work on the part of the predominant landowning groups is the main cause for large incidence of sharecropping in the areas of wet paddy cultivation.[29] While accepting Beteille's stand, we argue that the caste background of landlords is also an important factor causing a large incidence of tenancy in a paddy producing region. We hypothesise, therefore, that tenancy is a dominant feature of agriculture in a paddy growing region where the bulk of landlords belong to the castes which are not interested in the cultivation of land. In other words, the dominance of non-agricultural landlords is the basic cause of tenancy in a region of paddy cultivation.

The factors that account for a high level of tenancy in a paddy growing area would determine its form also. Hanumantha Rao has, in an important article, tried to explain the form of tenancy (fixed-rent *vis-a-vis* crop-sharing) in terms of the scope for decision-making and the magnitude of uncertainty faced in it. According to him, fixed cash rents are common where uncertainty is high and the scope for decision-making, for example, for factor and product substitution, is significant; but crop-sharing is common in areas of relative economic certainty with very little scope for decision-making or for entrepreneurial profits. He also adds that as profitable modern inputs assume significance, the incentives for capturing returns from increased investment may lead to preference for fixed cash rents and the cultivation of land with hired labour.[30] Set within a neo-classical framework, his analysis does not include the monopoly power of the landlord as a variable in determining the forms and terms of tenancy. Thus, who will decide if crop-sharing is to go, and if it is to be replaced by fixed cash rent or owner-cultivation? At one stage, he hints that tenants have a say,[31] but this is not consistent with his conclusion that crop-sharing can be replaced by owner-cultivation, unless he means that the owner-cultivation is brought about by tenant himself becoming the owner. But this is not ordinarily possible except through organised and sustained struggles by peasants against landlords.

Reverting to Beteille's exposition, it may be observed that even the work of personally supervising paddy cultivation in the monsoon season is something which many landowners prefer to avoid.[32] In a sharecropping arrangement, the landlord may be expected to involve himself in agriculture at least by way of supervision. But by leasing out land on fixed-rent basis, he can relieve himself from the work of supervision even. Therefore, we hypothesise that in a paddy producing region characterised by the preponderance of non-agricultural absentee landlords, fixed-rent tenancy is predominant.

It is also necessary to examine the causes for the growth of tenancy. Here, the consequences of landlord-controlled usury have to be analysed. Quite a few scholars in India have brought to the fore the significance of the usurious control of the landlord as a mode of exploitation. One of the important features of semi-feudalism in agriculture, as stated by Bhaduri, is the concentration of two modes of exploitation, namely, usury and landownership in the same economic class.[33] Prasad, lending support to Bhaduri's thesis, states that usury is practised mainly to perpetuate the indissoluble bond between the tenant and his overlord.[34]

The critics of Bhaduri's thesis, namely, Bardhan and Rudra, found that in the villages of West Bengal, Bihar and Uttar Pradesh, the incidence of landlord-controlled usury as a mode of exploitation was very less.[35] But the study undertaken by Harris in West Bengal itself disproves the validity of their finding. He found that the large body of poor peasants was still under the grip of landlords-cum-usurers in Sahajapur in 1982.[36] Bringing out the implications of interlocking of land, credit, labour and commodity markets, Krishna Bharadwaj has shown how a class, dominant in one market, dominated in others and this secured greater gains than if its operations were limited to one market alone.[37] Pani also shows that the small tenants and the landless labourers were under the usurious control of the relatively large landholders in southern Karnataka in the 18th and 19th centuries.[38]

Thus, there is concrete historical evidence to prove the significance of landlord-controlled usury as a mode of exploitation. It is not then difficult to see that this itself led to the growth of agricultural tenancy. In his analysis of the formation of the intermediary socio-economic structures in India in the colonial period, Rastyannikov shows how the small peasant indebted to the usurer was compelled to transfer his property title to the latter and subsequently became a tenant on the same land which he owned before.[39] We hypothesise here that in a region where the non-agricultural landlords acquire, in course of time, prominence as usurers also, the extent of tenancy increases due to the large scale transfer of property titles of the indebted owner-cultivators to the usurers, the former subsequently becoming the latter's tenants.

The flow of the merchant's capital into the countryside is also likely to alter the agrarian structure. Lenin shows through a wealth of statistics that the independent development of the merchant's and usurer's capital can retard the development of agrarian capitalism.[40] The distinct differentiation within the class of landed proprietors in Bengal which began by the middle of the nineteenth century, was mainly due to the entry of the trader's capital into agriculture. The landed proprietors in the early 19th century Bengal were still big zamindars with large estates. By the 1870s, many Bengali middle class people who had accumulated savings mainly through petty trade or commerce in the urban areas started buying proprietory or intermediary rights in land and eventually emerged as a dominant rentier class.[41] Keeping in view this historical reality, our hypothesis is that in a region where non-farm investment opportunities are few and the purchase of land is almost the only profitable mode of investing money, the penetration of the merchant's capital into the countryside eventually leads to the growth of parasitic absentee landownership.

The extent of tenancy, as already noted, may not be the same everywhere. Hence, the need for an analysis of factors accounting for regional variations in the extent of tenancy. Bardhan attempts to relate regional variations in the extent of tenancy at the macro-level, i.e., state level, to differences in (a) intensity of irrigation, (b) fluctuations of rainfall, (c) labour intensity of crops, (d) degree of landlord's monopoly power in the land market, (e) unemployment of landless labourers, and (f) imperfections in credit market.[42] Rudra, however, contends that the state is rather too large a unit for the analysis of regional variations in the extent of tenancy, for the variations inside any state are very large.[43] A study conducted by Beteille in Tanjore district vividly shows that variations in the extent of tenancy can exist between different regions of the same district.[44] He reports that tenancy as a mode of productive organisation is more common in the Old Delta region of Tanjore marked by fertile soil, better water supply, greater density of population and predominance of paddy cultivation than in the New Delta region where the natural factors are less favourable, the density of population is also less and dry crops are raised. We argue that in addition to the difference in the natural factors, the difference in the character of landlordism would also account for regional variations in the extent of tenancy in a district. Hence, we raise the following hypotheses in this regard:

(i) The extent of tenancy is more in the region of a district characterised by favourable natural factors and greater density of population than in the other region of the same district where the natural factors are unfavourable for human habitation.

(ii) The extent of tenancy is more in the region of a district where the non-agricultural absentee landlords are predominant than in the other region of the same district where the bulk of the landowners belong to the castes interested in agriculture.

The information regarding the magnitude of tenancy would not be very illuminating unless details about its character or distribution of terms and conditions in different forms of tenancy are known.[45] Myrdal made a distinction between the privileged tenants enjoying the hereditary rights over the leased land, fixity of rent and high degree of security and tenants without these privileges. He further points out that this demarcation between the privileged and unprivileged tenants caused surprise to the Europeans, though for Indians it was a logical extension of the ancient tenure arrangements and the caste system.[46] This demarcation reflects the relative bargaining strength of the two parties. A privileged tenant has bargaining strength *vis-a-vis* his landlord; but an unprivileged tenant is weak. To get an insight into the bargaining strength of the parties, we have to study the terms and distribution of benefits in different forms of tenancy.

The terms of tenancy contract in respect of duration, however, would be different for different crops. We hypothesise that permanent lease contract is suitable for those of the commercial crops the cultivation of which requires constant care and heavy long-term investment by the tenant. On the contrary, short-term or temporary lease contracts are likely to be predominant in the case of seasonal crops like paddy, if its cultivation does not necessitate any significant long-term investment. The rent payable by the permanent tenant would depend on the share of each party in investment and maintenance of the leased land and also on the expectations of the parties at the time of the execution of the lease contract. The greater the share of the landlord in investment and maintenance, the greater the rent and *vice versa*. In the period of high prices, the rent agreed upon by the tenant may be high as he would develop an optimistic forecast. The rent payable by the temporary tenant cultivating the paddy land would depend mainly on its fertility---greater the fertility of the soil, the greater being the rent. The modes of payment of rent would also depend on the nature of the crop raised on the leased land. Cash rents are likely to be predominant in the case of commercial crops, while kind rents are likely to be common in the case of food crops.

If tenancy is the direct outcome of usury, the bargaining capacity of the tenant would be very weak. If a peasant owing debt to a usurer sells the mortgaged land and takes it on lease, then it becomes interest-lease, for interest would be disguised as rent. In such cases, the debtor-cum-tenant is at the losing end and the creditor-cum-landlord can practically impose his terms on the former. In a traditional land-based social order, the possibility of feudal elements characterising the contractual terms cannot be completely ruled out. The tenant, in some cases at least, in addition to paying rent for the use of land, would be bound by contract to render unpaid labour and to make payments in kind just as a token of honour and recognition for the landlord. Such additional stipulations marked by the presence of extra economic coercion are feudal in character.

The caste factor is also likely to influence the bargaining strength of the tenant. If the two parties, lessor and the lessee, belong to the same caste, the latter is likely to secure equitable contractual terms. On the contrary, if the landlord belonging to the so-called superior caste leases out land to a tenant belonging to the so-called inferior caste, the contractual terms are likely to be unfavourable for the latter.

In extreme cases, even a landlord preoccupied with non-farm occupations may evince interest in the cultivation of paddy land. If, due to special ecological conditions surrounding the leased land, heavy investment and costly reclamation measures are necessary every year which the tenant cannot afford, the contractual terms and would be such that the tenant would contribute labour, and the landlord, without of course, participating in any of the agricultural operations, would invest his capital only if he is sure of getting adequate returns on it.

Consequent upon the regional variations in the extent of tenancy, there could be regional variations in the terms of tenancy also. In the region of a district where tenancy is dominant, the tenants would have weak bargaining power *vis-a-vis* the landlords as long as the former are not organised. On the other hand, in the other region of the same district characterised by lower incidence of tenancy, even if tenants are not organised, their bargaining strength can be significant if that region has a scarcity of labour and agricultural operations are more arduous even if profitable. Farm wages are obviously higher in such regions. The bargaining strength of the tenant thus depends on the farm wage because the alternative for tenancy cultivation for a non-cultivating landlord is cultivation through hired labour. Bardhan and Srinivasan argue that a region with a higher wage rate will tend to have a larger incidence of sharecropping tenancy.[47] But we hypothesise that in a region with a higher wage rate, the tenant can secure equitable contractual terms. On the contrary, in a region where farm wages are low, the terms for the tenant are unfavourable.

Followers of the neo-classical theory seem to take the affluence of the fixed-rent tenant for granted. Adams and Rask actually state that the fixed-rent tenants, being relatively affluent, are not as susceptible to exploitation as the tenants under traditional sharecropping system, i.e., sharecropping arrangement without cost-sharing by the landlord.[48] However, Myrdal's finding that the fixed-rent tenants with short-term or purely temporary lease contracts are as vulnerable as the share croppers clearly disproves their contention.[49] Hanumantha Rao's contention that conversion of crop-sharing rentals into fixed cash rents may be possible only in those situations where the tenants have a stronger bargaining power,[50] is contradicted by the findings of the study conducted by Krishna Bharadwaj and P.K. Das in Orissa.[51] They noted that, with the introduction of high yielding varieties of paddy in the areas of assured water supply, the landlords found it profitable to switch over to fixed-rent in kind or cash, for it could provide an incentive to the tenant to cultivate intensively since what remained after paying rent accrued to him. However, by making the lease contract short enough and/or insecure, the landlord could raise fixed-rent to capitalise on the productivity gains for each new tenant or for the old tenant under the threat of eviction. So, there is every reason to believe that unless contracts are for a considerably long period, the system of fixed-rent can be as exploitative as the traditional sharecropping system. It is quite possible that if fixed-rent and sharecropping systems exist side by side in respect of the same crop in the same region, the gains shared by tenants under both the systems would be more or less the same, though terms apparently differ.

There is an intellectual tradition of treating tenancy as a

semi-feudal or feudal phenomenon. As already noted, Bhaduri and Prasad regarded it as a semi-feudal arrangement of production in agriculture. In the course of his analysis of the features of over-populated agriculture in South-Asian countries, Dandekar also considered it essentially as a feudal phenomenon.[52] Rudra, however, took a different stand in this respect and asserted that tenancy arrangements could be straight-forward contracts between two parties and in such cases, the arrangement could fit very well into a capitalist social organisation or be a vehicle for emerging capitalist tendencies.[53] Aparajita Chakraborty who did not find anything in the writings of Marx or Lenin that equated tenancy with any pre-capitalist mode of production in a simplistic and mechanical manner, has emphasised the flexible character of tenancy by stating that it was compatible with both pre-capitalist and capitalist modes.[54] What happened in American agriculture during 1910-1964 indeed provides a grand proof to the flexible nature of tenancy. During this period, various semi-feudal forms of tenancy were ousted and there was development of capitalist tenancy on their bones.[55] In some villages of East India also, sharecropping tenancy with cost-sharing by landlords facilitated development of capitalism in agriculture.[56] These instances show that far from being rigid, tenancy can take different forms under different historical conditions. Besides, it is possible that tenancy, even if characterised by the presence of feudal features, can be an instrument of capitalist development in agriculture if its form and terms are basically conducive for such development. It may not be unrealistic to visualise the co-existence of both the feudal and capitalist modes of production in the same region and on account of its flexible character, tenancy may characterise both these modes. Given the flexible character of tenancy, it is important to see how the feudal mode of farming gets dismantled and the need for tenancy reforms arises.

The Working of Tenancy Reforms

History is replete with examples to show that a radical tenancy reform legislation, far from being a liberal gift from the government, is the outcome of a long and protracted struggle by the peasants.[57] In the first instance, the tenancy movement raises the bargaining power of the tenant *vis-a-vis* the exploitative landlord and thus initiates a process of change in the tenant-landlord relationship. Further, it brings about a sharp rise in the consciousness of tenants which is a prerequisite to the introduction of any legislative-administrative action for conferring rights and privileges. Joshi emphasises the importance of organisation and politicisation of peasants when he states that there are states like Kerala where a stronger left movement and greater degree of consciousness and organisation of peasants has not only ensured a better land reform legislation, but also more effective implementation of this legislation, while in states like Bihar, Madhya Pradesh and Rajasthan, the greater influence of the landlord at the political level and lack of consciousness and organisation of peasants has been responsible for landlord-oriented legislation and also its implementation in favour of landlords.[58]

In one of their recent papers, Suri and Raghavulu point out that in Andhra also the tenancy movement (movement of tenants-at-will under landowners) was never very strong. The problem of tenants-at-will was never seriously addressed even in the course of militant anti-zamindari and jagirdari struggles in the past. Therefore, the introduction of tenancy legislations in the Andhra area had only a negative impact on tenants, landowners resorting to large-scale evictions and leasing out lands only on an oral basis.[59] As Parthasarathy and Pothana argued, the 1960s and 1970s witnessed a rapid decline in the number of tenants and area under them, not due to change of tenants into owners but as a result of dispossession.[60] In recent years (1980s and early 1990s) the phase of reforms in Andhra Pradesh lost momentum due to the relative absence of pressure from the grassroots level and weaknesses of the peasant movements. The focus of peasant associations has shifted from the central one 'land-to-the-tiller' to issues such as remunerative prices, irrigation, subsidies, etc. In other words, the pure or temporary tenants in Andhra have not been able to get the reform that they required because of the absence of strong tenancy movements.

The initial stimulus for tenancy movement in a region may emanate from external sources; the external stimulus, however strong, cannot spark off tenancy movement unless the drawbacks in the land tenancy system necessitate a change. Zagoria,[61] in his research about peasant communism in South Asia, noted that the contradictions between landlord and tenant are evident in the paddy growing regions which constitute one of the important factors facilitating politicisation of peasants. Tenancy movement is thus an outcome of the contradictions in the land tenancy system.

Though the peasant struggles, by and large, have occurred largely due to economic causes, these struggles cannot be explained entirely in economic terms as is believed by many. No doubt, the basic grievances of the peasants may be economic, whether against landlords or feudal lords, moneylenders, jagirdars or the state itself. Zagoria[62] has proposed a set of social, political, psychological and economic conditions which produce the upsurge of the peasant struggles in South Asian countries, especially those South Asian countries producing rice. Heavy demographic pressure on land, growing class of landless labourers, exploitative land tenant system are some of the important economic causes leading to agrarian unrest in the paddy producing Asian countries. But there are also non-economic factors, mostly political and social, which supplement the economic factors. As stated by Zagoria himself, awareness of the possibility of change and the accessibility of the rural power structure, freedom to form organisations and indulge in militant activities and weakening of the strength of conservative landed elites and organisations, and manipulative skill for organising peasantry along the militant lines exhibited by their leaders are the significant non-economic factors. He further states that the poor must be aware of the power of their numbers and must be prepared to make it difficult for their class enemies to suppress them. Thus, from the vivid account of agrarian structure in the South Asian countries given by Zagoria, it is clear that there should be a combination of circumstances for the emergence of radical peasant movements and their growth.

Contrary to Zagoria's argument, some western writers who have examined the background of peasant participation in a revolutionary warfare, in Mexico, Russia and China have expressed doubt about the capacity of poor peasants to revolt. For instance, Wolf, who magnifies the revolutionary zeal of the so-called 'middle peasant', states that since the poor peasants and the rural proletariat have no tactical power, they are unlikely to pursue a rebellious course unless they have an external power to rely on.[63] While Alavi also speaks of the capacity of the middle peasant to rebel, he does not say that the poor peasant has no revolutionary potential at all. He writes that the poor peasants and rural proletariat form potentially the most revolutionary categories, but, because of their economic dependence, they are incapable of taking the initiative to begin a movement.[64] We argue, however, that the poor peasants, when guided by able and inspiring leaders, can pursue a rebellious course and stand to gain.

The role of leadership factor in peasant movements is also an important issue to be examined. The Subaltern School of thought on peasant movements generally takes the stand that the leadership factor does not assume much importance in the origin and growth of peasant movement in India. Hardiman,[65] who may be taken as a representative of the school, examines the role of leaders and political parties in his important paper, "The Indian Faction: A Political Theory Examined". He quotes Carter with approval who argues that there is a political class in India. This class consists of men with local influence and powers of patronage, such as the leaders of the dominant caste in a village, big landlords, and in the towns the moneylenders and capitalists. Whenever this class participates in election politics, it only seeks to get the support of the poor peasants by expressing lip sympathy. In other words, the leaders hardly play a useful role in organising peasants on militant lines. They are just interested in getting their personal interest fulfilled. Therefore, the grievances of the poor peasants are casually attended to by such leaders. On this count, the Subaltern School does not admit the significance of leadership factor in peasant movements. But, Sabha, in the light of the Tribal Peasant Movement in north-east India (Darjeeling and Sikkim), states that the grievances of the poor peasants alone cannot result in a revolt. To quote him: "The leadership, whatever may be the personal interests of the leaders themselves, has been found to be important. Above all, the political mobilisation of the peasantry, whether by individual leaders or political parties, has been found a necessary factor in any peasant movement. However, influential it may otherwise have been in the recent years, this conclusion may seem to contradict the Subaltern School of thought on peasant movements, but it has the strong support of the incidents in Sikkim."[66]

In the Malabar region of Kerala, in the 1920s and 1930s, the agrarian discontent became a potent force of change because it was properly channelised by able leaders.[67] Similarly, the Bihar Provincial Kisan Sabha, which came into existence on 29 November, 1929, under the charismatic leadership of Swami Sahajanand Saraswati, within a short time, could create sufficient pressure through meetings and publicity so that the government was forced to withdraw the proposed amendment to the Bihar Tenancy Act of 1885, which, if carried out, would have further weakened the position of tenants.[68]

The specific tasks which the leaders of the poor peasants generally perform are listed by Seth as under:[69]

1. to promote consciousness for group action among peasants to protect their rights and improve their bargaining power;
2. to help them to organise;
3. to manage the affairs of the organisation from day to day; and
4. in the event of a conflict with the rural elites, to provide support in facing the elites.

The problem of small landlords has, in the context of tenancy abolition, received special attention. For instance, the Owner Farmer Establishment Special Measures Law(OFESML) implemented in Japan in the post-war period, which is hailed as one of the most revolutionary land legislations, permitted the resident landlords there to lease out upto one hectare of land. When peasant unions opposed this provision, the Japan government justified it on the ground that it was necessary to help the small landowners who were not landlords in the socio-economic sense of the word but who, for some reason, were not able to cultivate the land personally.[70] The Land-to-the-Tiller Act passed in 1953 in Taiwan, another revolutionary legislation, also permitted small landlords to lease out upto three hectares of land.[71] In Kerala, where both landlords and tenants were poor, the land legislation had to perform a delicate function of balancing the interests of one weak party against that of a weaker party.[72] Thus, all such instances prove the vulnerability of the small landlord to any sweeping tenancy reform programme as well as to tenancy movement.

In the light of quite a few instances obtained in India, we can argue that tenancy legislation can produce the desired change if it affects the class of absentee landlords. Gadgil, in his preface to Dandekar's report, points out that the Bombay Tenancy and Agricultural Lands Act of 1948 for protecting tenants had greater effect in the Konkan district of Maharashtra because large number of Konkani landlords were absentees habitually residing in towns and cities. In the Desh district, on the other hand, the landlords, most of whom formed of the better-to-do among the body of peasants themselves, succeeded in getting voluntary surrender of leaseholds in terms of the Act. This happened even though the tenant in the Konkan district was poorer and traditionally much more subservient to the landlord than in the Desh district.[73] The Maharashtra Land Reforms Review Committee, in its report, stresses the fact that the aim of the Bombay Tenancy and Agricultural Lands (Amendment) Act of 1955 to confer ownership on tenants was fulfilled to a great extent in the coastal region of western Maharashtra where the majority of landlords were absentee landlords, whereas in the non-coastal region characterised by predominance of cultivating resident landowners, it was very much defeated.[74] We argue, therefore, that both tenancy movement and legislation have a greater chance of success in a region characterised by the predominance of small and non-cultivating absentee landlords.

As a corollary to our hypothesis of regional variations in the extent and terms of tenancy, we raise the hypothesis that such variations, besides causing localisation of tenancy movement, would account for the varying impact of the tenancy legislation. Furthermore, the benefits of the land-to-the-tiller legislation would be different for different categories of tenants. While the conferment of ownership would result in considerable gains for the tenant facing unfavourable contractual terms, it would be of little or no significance in the case of privileged tenants.

The importance of up-to-date and clean records of rights in the context of implementing the tenancy reform needs hardly any elaboration. One of the reasons for the non-implementation of many tenancy reform laws, particularly in the former intermediary tenure areas in India, is the absence of up-to-date land records. In order to confer or protect the legal rights of the tenants, it is necessary to have records relating to the ownership and tenancy on each plot of land involved in tenancy and the changes in this from year to year. There are no such records in most parts of the ex-zamindari areas in India except West Bengal even today, some 40 years after the abolition of zamindari. Consequently, it was not possible or easy to implement tenancy reform laws. In view of this fact, it may be stated that existence of up-to-date and clean land records facilitates the implementation of tenancy reform. The greater awareness of the tenants about their rights which is the outcome of the tenancy movement would enable them to use it at the right moment, including insistence on creation of right land records.

In the light of the foregoing discussions, this study examines the following hypotheses:

1. Tenancy is a dominant feature of agriculture in a paddy growing region where the bulk of landlords belong to the castes which are not interested in the cultivation of land.
2. In a paddy producing region characterised by the preponderance of non-agricultural absentee landlords, fixed-rent tenancy is predominant.
3. The extent of tenancy is more in the region of a district where the non-agricultural absentee landlords are

predominant than in the other region of the same district where the bulk of the landowners belong to the castes interested in agriculture.

4. The permanent lease contract is suitable for those of the commercial crops, the cultivation of which requires constant care and heavy long-term investment by the tenant, while short-term or temporary lease contracts are likely to be predominant in the case of seasonal crops like paddy, if its cultivation does not necessitate any significant long-term investment.
5. Consequent upon the regional variations in the extent of tenancy, there could be regional variations in the terms of tenancy also, which in turn, indicate the regional variations in the bargaining strength of the tenant *vis-a-vis* the landlord.
6. If fixed rent and sharecropping systems exist side by side in respect of the same crop in the same region, the gains shared by tenants under both the systems would be more or less the same though terms apparently differ.
7. The success of tenancy reform is the result of an increase in the awareness of tenants brought about by the tenancy movement.
8. The varying impact of tenancy reform and localisation of tenancy movement are the consequences of regional variations in the extent and terms of tenancy.
9. Poor peasants, when led by dynamic leaders, can take to a rebellious course and achieve the benefits of tenancy reform.
10. Both the tenancy movement and reform have a greater chance of success in a region characterised by the predominance of small and non-agricultural absentee landlords.

Methodology

Our study needed an interdisciplinary approach combining economics, sociology and political science. It obviously required a historical treatment, involving a study of how agrarian structure evolved in Uttara Kannada district, what contradictions it contained and how tenancy movements and legislations sought to resolve these contradictions. This makes our approach close to what is described above as one of the political economy in so far as the objectives are viewed in historical institutional, interdisciplinary and above all, a class perspective. At the same time, we have not ignored ecological, physical and technical factors that had a bearing on the distribution of terms and conditions in different forms of tenancy that existed in the district.

This study is based on only secondary data including historic documents and interviews with leaders of peasant movements in Uttara Kannada. Though the Census of Agricultural Holdings, conducted in 1970-71, indicates the magnitude of tenancy and its regional variation, it is quite likely to be underestimated due to the provision for resumption of land under personal cultivation and dispossession of tenants under the provision of voluntary surrender allowed in the earlier laws. Besides, the census does not provide a comprehensive picture of the district. In these circumstances, it becomes essential to use other forms of information which would supplement the hypotheses arising out of secondary data. In this connection, information provided by historical documents could play a crucial role. The important historical documents used by us are the Abstracts of Correspondence between the Madras Board of Revenue and the Collectors of Kanara, the Kanara Gazetteer published by the Government of Bombay and the Land Revenue Settlement Reports relating to the different talukas of the district published during the pre-independence period.

For obtaining authentic information on the origin, growth and decline of tenancy movement, we have referred to the leaflets published by various peasant associations, the relevant issues of the local Kannada Weeklies, namely, *Kanada Vritta* (Kanara News) and *Janasevaka* and also booklets containing debates in Bombay Legislative Assembly and Council. Besides, our interviews with leaders, namely, Dinakar Desai, Dayanand Nadkarni and S.V. Pikle who played a vital role as peasant leaders, were instrumental in revealing the class character of the tenancy movement. We have consulted books that contain vivid discussions of the different tenancy legislations that operated in the district. In the light of the official data, we have attempted to bring out the impact of the Karnataka Land Reforms Act of 1974.

Overview of the Book

With a view to introduce our study area, we present in the section that follows a brief picture of the physical features, land utilisation pattern, cropping pattern, size structure of operational holdings, taluka-wise density of population in, and also the historical background of Uttara Kannada. Chapter II tries to bring out the causes of tenancy, the regional variations in the extent of tenancy and the factors that influenced the distribution of terms and benefits between the two parties in different forms of tenancy that existed in Uttara Kannada. Besides, it brings to light the contradictions in the land tenancy system that led to the outbreak of the peasant movement in the district in 1940. Chapter III describes how the tenancy movement and legislations sought to resolve the contradictions in the land tenancy system in Uttara Kannada in the pre-reorganisation period. Chapter IV discusses the impact of the peasant movements and legislations in the district during the post-reorganisation period. Chapter V contains summary and conclusions.

The Study Area and its Agricultural Economy

Uttara Kannada, also called North Kanara, with Karwar as its headquarters, is the northern coastal Malnad district of Karnataka State. It lies between 13°55' and 15°31' North latitude and between 74°9' and 75°10' East longitude. The district is bounded on the north by Goa and Belgaum district, on the south by Shimoga and Dakshina Kannada districts and on the west by the Arabian Sea (see Map 1).

Uttara Kannada has an area of 10,291 sq.kms., which forms nearly 5.4 per cent of Karnataka. Even now, it is known as a forest district for, according to the official data presented in Table 1.2, forest area accounts for 81 per cent of geographical area, while net cropped area is around 11 per cent. A somewhat broken and irregular range of Sahyadri hills divides the district into two distinct geographical tracts, the above-ghat or Malnad region with an area of more than 7,676 sq.kms., and the below-ghat or coastal region which covers an area of more than 2,404 sq.kms. Six talukas of Haliyal, Mundagod, Siddapur, Sirsi, Supa and Yellapur constitute the Malnad region, while Ankola, Bhatkal, Honavar, Karwar and Kumta talukas form the coastal region (see Map 2). The coastal region too has a significant part of area under forests (see Table 1.2). The most striking feature of the below-ghat talukas is the sharp distinction that exists between narrow coastal tract and the very much more extensive inland tract of hilly and mountainous formation.

The climate of the district is generally hot along the coast and cool in the up-ghat region. Uttara Kannada is indeed one of the few blessed districts in India which has hardly ever faced a drought, for the rainfall here is certain and plentiful, its normal average being 2,764 mm.

Map 2
Uttara Kannada: Taluka Boundaries

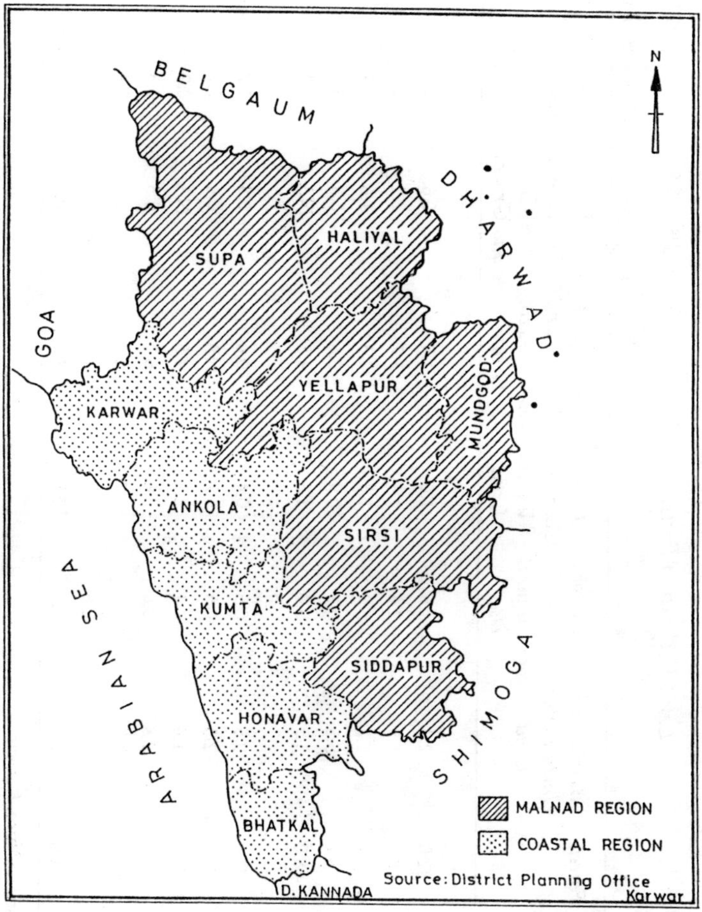

Table 1.2

Land Utilisation Pattern in Uttara Kannada, 1994-95

(figures as % of total geographical area)

S. No.	Taluka	Forests	Barren and Unclutivable Land	Land for Non-agricultural Uses	Other Cultivable Waste Land	Permanent Pastures and Other Grazing Groves	Land under Miscellaneous Tree Crops and Groves	Current Fallows	Other Fallow Land	Net Sown Area
Coastal Regions										
1.	Ankola	82.0	2.3	1.6	1.0	1.5	0.6	0.6	1.2	9.2
2.	Bhatkal	68.9	1.8	2.6	4.4	3.7	2.2	0.2	0.1	16.1
3.	Honavar	74.4	2.7	5.8	1.3	2.3	1.6	0.4	0.5	11.0
4.	Karwar	75.7	5.5	0.5	0.9	0.6	0.8	3.6	2.1	11.2
5.	Kumta	68.0	3.9	5.0	1.6	4.3	NA	0.4	1.2	15.6
Malnad Regions										
6.	Haliyal	70.6	2.5	0.5	0.4	0.6	0.8	0.2	0.3	24.3
7.	Mundagod	75.4	2.3	0.5	1.7	NA	0.6	0.2	0.3	18.0
8.	Siddapur	76.2	1.6	1.4	2.5	4.5	NA	0.3	0.7	12.8
9.	Sirsi	78.4	2.8	0.6	0.6	6.2	0.5	0.3	0.4	10.8
10.	Supa (Joida)	92.2	0.7	1.6	0.7	0.2	0.4	0.4	0.9	2.9
11.	Yellapur	90.5	0.7	0.6	0.2	1.0	NA	0.9	0.5	5.6
	District	81.0	2.0	1.4	0.8	2.1	0.5	0.6	0.7	10.9

Source: District Statistical Office, Karwar.

The principal rivers in the district are Kali, Gangavali, Aghanashini, Sharavathi, Venkatapur and Varada. Except for the Varada which flows eastwards, all other rivers flow east to west, cutting the Western Ghats into deep valleys, thus rendering them almost unfit for irrigation. The highly limited scope for irrigation in Uttara Kannada is something which needs to be recognised while assessing the role of tenancy reforms in agricultural development at present.

Table 1.3

Percentage of Area Under Important Crops to Gross Cropped Area in Uttara Kannada, 1994-95

S No.	Taluka	Paddy	Other Cereals	Pulses	Sugarcane	Groundnut	Coconut	Arecanut	Other Crops
Coastal Region									
1.	Ankola	67.1	3.2	0.3	0.6	12.6	6.3	2.3	7.6
2.	Bhatkal	62.1	4.2	1.1	0.9	15.3	8.6	3.1	4.7
3.	Honavar	55.6	5.5	0.3	1.9	9.5	14.7	8.8	3.9
4.	Karwar	80.4	0.9	0.1	0.2	2.4	8.5	0.3	7.2
5.	Kumta	66.0	2.3	0.3	1.2	8.9	9.8	4.1	7.4
Malnad Region									
6.	Haliyal	65.5	13.5	1.8	2.8	1.2	0.1	-	14.9
7.	Mundagod	74.2	8.0	2.3	0.3	0.8	0.4	0.2	13.9
8.	Siddapur	57.4	8.0	1.3	1.6	0.7	0.3	18.9	11.8
9.	Sirsi	69.7	1.1	2.5	1.8	1.1	1.2	16.9	8.3
10.	Supa (Joida)	86.5	2.1	0.5	4.3	0.1	1.2	1.4	3.9
11.	Yellapur	69.8	3.7	0.3	1.1	0.4	3.5	13.5	7.7
	District	66.8	5.4	1.1	1.6	4.2	4.2	6.4	9.9

Note: Only important crops are included.

Source: District Statistical Office, Karwar.

The soil is mostly lateritic and acidic in the talukas of Honavar, Kumta, Bhatkal and parts of Supa. In Karwar and Ankola talukas laterite mixed with gravelly soil derived from granites and sandy soil are found. Along the coast, patches of red loam predominates. However, patches of alluvial soil locally called *gadde* are also found. Near the mouth of the river valleys, a black loose marshy land called *gazni* occurs. Soil in the upper slopes of the valleys locally called *makki* consists of very hard laterite, which, if not constantly worked, stiffens into clods. The soil in Sirsi, Siddapur and Yellapur is red laterite *kagdale*, with average fertility which is very well suited for garden crops. Red loam derived from laterite and traps and patches of black soil are noticed in Mundagod, Haliyal and Supa talukas.

Table 1.3 shows the important crops grown in Uttara Kannada. Paddy is the principal food crop which accounts for as much as 67 per cent of the gross cropped area in the district. In recent years, the importance of groundnut as a rabi crop has been rising especially in the coastal region. Coconut is the chief garden crop in the coastal tract. Sirsi, Siddapur, Yellapur and the inland portions of Kumta and Honavar have a large area of betelnut gardens with pepper, cardamom and banana as inter-crops in the first three talukas.

As Table 1.4 shows the net irrigated area which constitutes just 22 per cent of the net sown area in Uttara Kannada. There is no major irrigation project in Uttara Kannada. The only major irrigation is provided by wells, tanks and *bhandaras* or canals. But it is only *bhandaras* and wells which supply water to summer crops as tanks generally go dry then.

The size structure of holdings in 1990-91 is presented in Table 1.5. The marginal holdings constituted more than 66 per cent of the total number of holdings in Uttara Kannada though they accounted for 21 per cent of the total operated area. While

Table 1.4

Irrigation Capacity by Different Sources in Uttara Kannada, 1994-95

S. No.	*Taluka*	*Wells*	*Tanks*	*Other Sources*	*% of Net Irrigated Area to Net Sown Area*
Coastal Region					
1.	Ankola	608	-	648	15.3
2.	Bhatkal	364	44	770	20.8
3.	Honavar	1,460	403	2,161	49.6
4.	Karwar	272	156	436	10.6
5.	Kumta	1,053	63	1,387	28.2
Malnad Region					
6.	Haliyal	-	4,218	317	23.5
7.	Mundagod	12	3,717	6	29.6
8.	Siddapur	567	268	750	14.1
9.	Sirsi	457	1,054	684	14.8
10.	Supa (Joida)	-	101	740	14.7
11.	Yellapur	167	287	1,385	22.9
	District	4,960	10,311	9,284	22.2

Note: The area irrigated through different sources is in hectares.

Source: Annual report for 1995-96 District Statistical Office, Karwar.

a large majority of holdings in the coastal talukas are exceedingly small, holdings in the up-ghat talukas are comparatively big.

The total population of the district, as per the 1991 census, was 12,20,260 (see Table 1.6). The density of population is higher in the below-ghat talukas than in the up-ghat talukas.

The history of Uttara Kannada[75] can be traced as far back as 240 BC, when it seems to have had associations with North India rather than with the South. In the 5th century AD, this district formed a part of the kingdom of Kadambas who had

Table 1.5
Size-Class-wise Proportions of Operational Holdings in Uttara Kannada, 1991

S. No.	Taluka	Marginal Holdings		Small Holdings		Semi-Medium Holdings		Medium Holdings		Large Holdings	
		No.	Area	No.	Area	No.	Area	No.	Area	No.	Area
Coastal Region											
1.	Ankola	76.6	31.1	14.2	26.8	7.1	25.9	1.9	14.4	0.1	1.9
2.	Bhatkal	81.3	38.6	12.8	28.5	4.7	20.2	1.2	10.6	0.1	2.2
3.	Honavar	83.5	37.9	10.3	23.8	4.6	21.0	1.5	14.2	0.1	3.0
4.	Karwar	84.2	43.6	11.2	28.1	3.1	18.4	0.7	7.6	0.1	3.3
5.	Kumta	80.6	35.9	12.7	27.6	5.3	22.2	1.4	11.8	0.1	2.4
Malnad Region											
6.	Haliyal	19.7	4.5	32.1	19.3	30.1	33.9	17.2	37.5	0.1	4.8
7.	Mundagod	18.8	5.9	45.6	31.1	24.6	31.7	10.4	25.9	0.7	5.4
8.	Siddapur	58.4	23.1	25.8	30.9	12.4	28.9	3.2	15.1	0.1	1.8
9.	Sirsi	44.1	12.8	27.6	22.2	18.9	29.6	8.4	27.2	1.0	8.1
10.	Supa	27.8	7.4	30.6	21.0	31.5	39.9	9.2	25.9	0.9	5.7
11.	Yellapur	41.3	14.2	31.1	28.1	21.7	36.9	5.9	19.9	0.1	0.6
	District	66.3	21.2	18.7	25.6	10.7	28.4	3.9	20.9	0.3	3.9

Source: District Statistical Office, Karwar.

their capital at Banavasi in Sirsi taluka of the district. During the subsequent period, the district came under the Vijayanagar kings from 1330 to 1580, and thereafter, under the Bijapur kings from 1580 to 1670. From 1672 to 1763, it was ruled by the local Sonda kings, with their capital near Sirsi. Later on, from 1763 to 1799, the district became subject to the rule of Haidar Ali and his son Tippu Sultan. After the Third Mysore War in 1799 and with the defeat of Tippu at the hands of the British,

Table 1.6
Population in Uttara Kannada, 1991

S. No.	Taluka	Density of Population per sq. km.	% of Rural Population to Total Population
1.	Ankola	99	85.2
2.	Bhatkal	370	75.6
3.	Haliyal	174	47.8
4.	Honavar	193	88.9
5.	Karwar	192	56.6
6.	Kumta	230	80.5
7.	Mundagod	112	100.0
8.	Siddapur	107	100.0
9.	Sirsi	116	66.7
10.	Supa	25	92.8
11.	Yellapur	51	76.9
	District	119	79.2

Source: Derived from 1990-91 census.

the present Uttara Kannada and Dakshina Kannada were formed into one collectorate of the Madras Presidency and Sir Thomas Munro was appointed as the Collector of Kanara by the British Company Government in June 1799. In 1862, Uttara Kannada was separated from Dakshina Kannada, and was transferred to the Bombay Presidency for purposes of administration. On reorganisation of states from November 1, 1956, it became a part of the erstwhile Mysore State (Karnataka).

The Bombay government initiated survey settlement operations in the district in the late 1860s and by 1891 lands were surveyed in all inhabited villages.[76] In the wake of survey settlement operations carried out from time to time, every holding of a specific size with a definite survey number came to be kept carefully distinct, and different rights in land were recorded. Subsequently, Maxwell was, in 1910, very much impressed by the remarkably efficient state of the records of rights.[77] Rajan, former Land Reforms Adviser to the Government of Karnataka, has also referred to the efficiency of land record maintenance in the Bombay-Karnataka region of which Uttara Kannada is obviously a part.[78]

Notes

1. James Manor referred to this view based on discussions with high officials in Karnataka and also with Devraj Urs; his own view regards this as the "most promising non-communist regional experiment". See James Manor, "Pragmatic Progressives in Regional Politics: Case of Devraj Urs", *EPW*, Vol. 15, Nos. 5, 6 and 7, Annual Number, February 1980, esp. pp. 201 and 205.
2. A study covering six villages of Marathwada region of Maharashtra showed, for example, that small farmers who have no resources may lease out land to large farmers. See M.V. Nadkarni, "Tenants from the Dominant Class: A Developing Contradiction in Land Reforms", *EPW*, Vol. 11, No. 51, Review of Agriculture, December 1976, pp. A-137-146.
3. Narendar Pani, *Reforms to Pre-empt Change: Land Legislation in Karnataka*, Indian Institute of Management, Bangalore, 1981 (mimeo.), p. 149.
4. M.A.S. Rajan, *Land Reforms in Karnataka: An Account by a Participant Observer*, Hindustan Publishing Corporation, Delhi, 1986, p. 109.
5. P.C. Joshi, *Land Reforms in India: Trends and Perspectives*, Allied Publishers, Bombay, 1975, p. 107.

6. Daniel Thorner, *Agrarian Prospect in India*, Delhi University Press, 1956, p. 8.
7. P.C. Joshi, *op.cit.*, p. 67.
8. Andre Beteille, *Studies in Agrarian Social Structure*, Oxford University Press, New Delhi, 1974, p. 25.
9. Michael Todaro, *Economic Development in the Third World*, Longman Group, New York, 1983, p. 519.
10. S.I. Cohen, *Agrarian Structures and Agrarian Reform*, Martinus Nijhoff Social Division, Leiden, 1978, p. 5.
11. Pranab Bardhan, "Variations in Extent and Forms of Agricultural Tenancy-I", *EPW*, Vol. 11, No. 38, September 18, 1976, p. 1545.
12. C.H. Hanumantha Rao, "Agricultural Growth and Stagnation", in A.M. Khusro (ed.), *Readings in Agricultural Development*, Allied Publishers, Bombay, 1968, p. 96.
13. Doreen Warriner, "Land Reform and Economic Development", in Carl Eicher and Lawrence Witt (eds.), *Agriculture in Economic Development*, Vora and Co., Bombay, 1964, p. 273.
14. V.M. Dandekar and Nilkantha Rath, "Poverty in India-II: Policies and Programmes", *EPW*, Vol. 6, No. 2, January 9, 1971, p. 111.
15. Alice Thorner, "Semi-Feudalism or Capitalism? Contemporary Debate on Classes and Modes of Production in India", *EPW*, Vol. 17, No. 49, (December 4, 1982), No. 50, (December 11, 1982) and No. 51 (December 18, 1982), pp. 1961-68, 1993-99 and 2061-66.
16. Utsa Patnaik, "Capitalist Development in Agriculture", *EPW*, Vol. 6, No. 39, Review of Agriculture, September 1971, pp. A-123-130.
17. Daniel Thorner, *Shaping of Modern India*, Allied Publishers, New Delhi, 1980, pp. 238-253.
18. Ashok Rudra, "In Search of the Capitalist Farmer", *EPW*, Vol. 5, No. 26, Review of Agriculture, June 1970, pp. A-85-87.
19. Bardhan speaks of five classes, with family farmers in the middle, who neither lease in nor lease out labour. Since such a

class hardly exists, we have adopted a more realistic scheme. See Pranab Bardhan, "Agrarian Class Formation", *The Journal of Peasant Studies*, Vol. 10, No. 1, October 1982, pp. 74-77.

20. Kathleen Gough, "Peasant Resistance and Revolt in South India", *Pacific Affairs*, Vol. 61, No. 4, Winter 1968-69, p. 527.
21. See, for instance, Doreen Warriner, *Land Reform in Principle and Practice*, Oxford University Press, Bombay, 1969, p. 141, P.S. Appu, "Agrarian Structure and Rural Development", *EPW*, Vol. 9, No. 39, Review of Agriculture, September 1974, p. A-71, and V.N. Reddy and C.S. Murthy, "Backward Castes and Tenancy", *EPW*, Vol. 13, No. 26, July 1, 1978, p. 1076.
22. Andre Beteille, *op.cit.*, p. 53.
23. D.W. Adams and N. Rask, "Economics of Cost Sharing Leases in the Less Developed Countries", *American Journal of Agricultural Economics*, Vol. 50, November 1968, p. 937.
24. A.M. Khusro, "Economic Theory and Indian Agricultural Policy", in A.M. Khusro (ed.) *op.cit.*, pp. 43-44.
25. A.J. Grad, *Land and Peasant in Japan*, International Secretariat, New York, 1952, p. 254. Also see *Agriculture in Japan*, F.A.O. Association, Tokyo, 1958, p. 5.
26. Masaru Kajita, *Land Reform in Japan*, Agricultural Development Series, Tokyo, 1959, pp. 10-11.
27. V.G. Rastyannikov, *Agrarian Evolution in a Multiform Structure Society: Experience of Independent India*, Routledge and Kegal Paul, London, 1981, pp. 252-253
28. C.H. Hanumantha Rao, *op.cit.*, pp. 94-96.
29. Andre Beteille, *op.cit.*, p. 29.
30. C.H. Hanumantha Rao, "Uncertainty, Entrepreneurship and Share Cropping in India", *Journal of Political Economy*, Vol. 79, June 1971, pp. 578-595.
31. He says, "The experience of a significant scope for entrepreneurship may occasion fixed cash rents, because the latter at the same time permit the *tenants* to capture the returns expected in consequence of their decision making....", *Ibid.*, p. 580 (emphasis added).
32. Andre Beteille, *Inequality and Social Change*, Oxford University

Press, Bombay, 1972, p. 27.

33. Amit Bhaduri, "Study in Agricultural Backwardness Under Semi-Feudalism", *The Economic Journal*, Vol. 83, March 1973, pp. 121-124.
34. Pradhan H. Prasad, "Reactionary Role of Usurers' Capital in Rural India", *EPW*, Vol. 9, Nos., 32, 33 and 34, Special Number, February 1978, p. 373.
35. Pranab Bardhan and Ashok Rudra, "Inter-linkage of Land, Labour and Credit Relations", *EPW*, Vol. 13, Nos. 6 and 7, Annual Number, February 1978, p. 373.
36. John Harris, *Making out on Limited Resources: or What Happened to Semi-Feudalism in a Bengal District*, University of East Anglia, Norwich.
37. Krishna Bharadwaj, *Production Conditions in Indian Agriculture: A Study Based on Farm Management Survey*, Cambridge University Press, London, 1974, pp. 3-5.
38. Narendar Pani, *op.cit.*, p. 139.
39. V.G. Rastyannikov, *op.cit.*, p. 139.
40. V.I. Lenin, *The Development of Capitalism in Russia*, Progress Publishers, Moscow, 1977 (Fifth Edition), p. 187.
41. Partha Chatterjee, "Agrarian Structure in Pre-partition Bengal", in A.K. Sen, Partha Chatterjee and Sagata Mukherjee (eds.), *Perspectives in Social Sciences-2*, Oxford University Press, Calcutta, 1982, pp. 119-120.
42. Bardhan concentrates mainly on the regional variations in the extent of sharecropping tenancy. Pranab Bardhan, "Variations in the Extent and Forms of Agricultural Tenancy-I", *EPW*, Vol. 11, No. 37, September 11, 1976, pp. 1507-1509.
43. Ashok Rudra, *Indian Agricultural Economics: Myths and Realities*, Allied Publishers, Bombay, 1982, p. 142.
44. Andre Beteille, 1974, *op.cit.*, pp. 152-156.
45. P.C. Joshi, *op.cit.*, p. 104.
46. Gunnar Myrdal, *The Asian Drama: An Inquiry into the Poverty of Nations*, Vol. 1, Allen Lane Penguin Press, London, 1968, p. 1038.
47. P.K. Bardhan and T.N. Srinivasan, "Crop-sharing Tenancy in

Agriculture: A Theoretical and Empirical Analysis", *American Economic Review*, Vol. 61, March 1971, p. 54.

48. D.W. Adams and N. Rask, *op.cit.*, pp. 940-941.
49. Gunnar Myrdal, *op.cit.*, p. 1066.
50. C.H. Hanumantha Rao, *Technological Change and Distribution of Gains in Indian Agriculture*, Macmillan Company, New Delhi, 1975, p. 154.
51. Krishna Bharadwaj and P.K. Das, "Tenurial Conditions and Mode of Exploitation", *EPW*, Vol. 10, Nos. 5, 6 and 7, Annual Number, February 1975, p. 237.
52. V.M. Dandekar, "Agricultural Growth with Social Justice in Over-populated Countries", *EPW*, Vol. 5, Nos. 29, 30 and 31, Special Number, July 1970, pp. 1233-1235.
53. Ashok Rudra, "Class Relations in Indian Agriculture-II", *EPW*, Vol. 12, No. 23, June 10, 1978, p. 964.
54. Aparajita Chakraborty, "Tenancy and Mode of Production", *EPW*, Vol. 16, No. 13, Review of Agriculture, March 1981, p. A-5.
55. V.G. Rastyannikov, *op.cit.*, p. 252.
56. Ashok Rudra, 1982, *op.cit.*, pp. 128-133.
57. P. Radhakrishnan, "Peasant Struggles and Land Reforms in Malabar", *EPW*, Vol. 15, No. 50, December 13. 1980, p. 2095.
58. P.C. Joshi, *op.cit.*, p. 96.
59. K.C. Suri and C.V. Raghavulu, "Agrarian Movements and Land Reforms", in B.N. Yugandhar (ed.), *Land Reforms in India: Andhra Pradesh: People's Pressure and Administrative Innovations*, Sage Publications, New Delhi, 1996, p. 44.
60. G. Parthasarathy and V. Pothana, "Changing Agrarian Structure in Andhra Pradesh", *Journal of Rural Development*, Vol. 2, No. 6, Nov. 1983, pp. 581-95.
61. Donald S. Zagoria, *"Communism in South Asia: Kerala and West Bengal"*, Problems of Communism, Vol. 22, 1973, Washington, pp. 16-27, quoted in Theodor Bergmann, *Agrarian Reforms in India*, Agricole Publishing Academy, New Delhi, 1984, p. 24.
62. Donald S. Zagoria, "Asian Tenancy Systems and Communist Mobilisation of Peasantry" The American Political Science

Review, 65, March 1971, pp. 144-160.

63. Eric Wolf, "On Peasant Rebellions" in T. Shanin (ed.), *Peasants and Peasant Societies*, Penguin Book, Harmondworth, 1969, p. 268.

64. Hamza Alavi, "Peasants and Revolution", in D.N. Dhanagare, *Peasant Movements in India (1920-50)*, Oxford University Press, New Delhi, 1983, p. 10.

65. David Hardiman, "The India Faction: A Political Theory Examined", in Rajnit Guha (ed.), Subaltern Studies I, Oxford University Press, Oxford, 1982, p. 219.

66. Tanka Bahadur Sabha, "Peasant Struggle in Darjeeling and Sikkim", M.N. Karna (ed.), *Peasant and Peasant Protests*, Intellectual Publishing House, 23, Daryaganj, Ansari Road, New Delhi, 1989, p. 127.

67. P. Radhakrishnan, "Peasant Struggle and Land Reforms in Malabar", *EPW*, Vol. No. 15, December 1980, p. 2097.

68. Arvind N. Das, "Peasant and Peasant Rebellions: The Kisan Sabha in Bihar", *The Journal of Peasant Studies*, Vol. 9, No. 3, April 1982, p. 59.

69. A.N. Seth, *Peasant Organisations in India*, B.R. Publishing Corporation, Delhi, 1984, p. 85.

70. Sidney Klein, *The Pattern of Land Tenure Reform in East Asia after World War II*, Bookman Monograph, New York, 1958, p. 30.

71. T.H. Shanin, "Land Reform and Agricultural Development in Taiwan", in J. Brown and S. Lin (eds.), *Land Reform in Developing Economies*, University of Hartford, 1968, p. 354.

72. P.C. Joshi, *op.cit.*, p. 104, Also see Theodor Bergmann, *op.cit.*, pp. 54-58.

73. See V.M. Dandekar and G.J. Khudanpur, *Working of Bombay Tenancy Act, 1948: Report of Investigation*, Gokhale Institute of Politics and Economics, Poona, 1957, p. iv.

74. Government of Maharashtra, Revenue and Forest Department, *The Report of the Committee Appointed for Evaluation of Land Reforms in Maharashtra*, Government Central Press, Bombay, 1974, pp. 74-78.

75. Superintendent of Census Operations, Census of India, 1961, Mysore State, *District Census Hand Book*, North Kanara District, Government Press, Bangalore, 1968, pp. iv-v.

76. *Imperial Gazetteer of India*, Provincial Series, Bombay Presidency, Vol. 2, Superintendent of Government Printing, Calcutta, 1909, p. 92.

77. R.M. Maxwell, *Papers relating to the First Revision Settlement of the Kumta Taluka of the Kanara Collectorate*, Government Central Press, Bombay, 1918, p. 17.

78. M.A.S. Rajan, "Working of Land Tribunals of India", in A.R. Rajapurohit (ed.), *Land Reforms in India*, Ashish Publishing House, New Delhi, 1984, p. 160.

2

Early Agrarian Structure and Factors Behind Tenancy in Uttara Kannaḍa

One of the distinguishing features of the early agrarian structure in Uttara Kannada was the inveteracy of the system of private property in land accompanied by the right of inheritance and unrestricted privilege to transfer it by gift, sale and lease, etc.[1] Munro, who was, in the beginning of his career the Collector of Kanara, inclined to recommend zamindari system of land revenue settlement, changed his mind later, and wrote to the Madras Board of Revenue that it was not worthwhile to initiate any process that would hamper the system of private property in land well-maintained in the region.[2] Therefore, the district remained a classic ryotwari area with almost total absence of any alienated holdings such as *inam* lands found in the neighbouring districts like Dharwad and Belgaum. Even lands attached to temples here were registered in the name of the deity; they were fully assessed so that there was no difference at all between the temple land and private holdings.[3]

Munro himself reported that tenancy was a dominant feature of agriculture in the district since centuries. In the thirteenth century itself, the persons holding land could be divided into three distinct groups, viz., Mulvargadars or original proprietors, Mulgenidars or permanent tenants and Chalgenidars or yearly tenants.[4] Francis Buchanan, who travelled in this part of the country and wrote his diary in 1801, noted that parts of holdings were being cultivated by tenants or *gaynicaras*.[5] This fact is further substantiated by correspondence between the Madras Board of Revenue and the Collectors of Kanara.[6] According to the 1872 census, 45.5 per cent of those engaged in agriculture in the district were tenants.[7] The Kanara Gazetteer also reported about the dominance of tenant labour.[8] The data presented in Table 2.1, besides indicating very high proportion of tenants and sharecroppers to total cultivators in Uttara Kannada, show that due to the dominance of tenant labour only, agricultural labour here was much less than in Karnataka as a whole in 1961 as well as in 1971. Thus, historical evidence and also recent secondary data consistently point towards very high level of tenancy in Uttara Kannada district.

High level of tenancy here could be attributed to the predominance of landlords belonging to the castes which were not interested in paddy cultivation which was, needless to state, labour intensive and arduous.[9] In the Malnad region, the Havik Brahmins, Goud Saraswat Brahmins, Saraswat Brahmins and the Lingayats were the chief landowning castes. The Haviks, numerically the most important landowning caste in this region, especially in Sirsi, Siddapur and Yellapur talukas, were primarily spice gardeners. The Kanara Gazetteer complimented them: "The owners of gardens are Haviks or Haig Brahmins who bring the labour from the coast and live in their gardens all the year. They are the best cultivators in Kanara and give the country its special character."[10] Indeed, the Haviks were capitalist farmers who employed wage labour on a large scale,

Table 2.1
Agricultural Workforce in Uttara Kannada and Karnataka

Year	*Male*			*Female*		
	A	*B*	*C*	*A*	*B*	*C*
Uttara Kannada						
1901	46.9	39.4	9.6	15.1	48.0	13.5
1911	48.9	NA	16.6	34.8	NA	36.6
1921	50.6	NA	14.0	35.1	NA	26.4
1931	54.6	80.2	8.0	31.0	83.6	33.3
1941	NA	NA	NA	NA	NA	NA
1951	52.6	57.9	11.8	54.3	14.3	21.2
1961	42.2	NA	9.6	54.0	NA	12.4
1971	38.1	NA	14.8	36.4	NA	29.1
Karnataka						
1961	53.4	NA	12.6	55.6	NA	23.7
1971	44.1	NA	21.2	24.6	NA	49.4

Notes: (i) A = Cultivators as % of total workers.
B = Tenants as % of cultivators.
C = Agricultural labourers as % of total workers.
(ii) The figures for agricultural labour from 1901 to 1951 refer to farm servants and field labourers together, including labourers in horticulture. In 1961 and 1971, agricultural labour is not inclusive of labour engaged in horticulture. The latter figures are not available at the district level; cultivators and tenants/sharecroppers were referred to only in crop husbandry.

Source: Census reports of India for various years.

produced marketable surplus and realised it through sale, for betelnut, pepper and cardamom grown by them enjoyed fame even in the world market.[11] But being prevented by rigid caste rules from holding the plough, they got their rice fields cultivated either through hired labour or by tenants. The Goud Saraswats and Saraswats took to government jobs and trade in towns and distant cities like Bombay and were least interested in agriculture so that Collins considered them as non-agricultural absentee landlords.[12] The Lingayats were a predominant land-owning caste in the Haliyal taluka who evinced keen interest in paddy cultivation. They were also found in the eastern part of Sirsi and in some villages of Mundagod taluka.[13]

In the coastal region, the dominant landowning castes were Goud Saraswats, Saraswats, Haviks, Sonars or Goldsmiths, Muslims, Nadors and the Konkan Marathas.[14] In contrast to the Malnad region, the Havikas were outnumbered by others in this region. Like their counterparts in the above-ghat region, the Haviks showed interest in garden cultivation in the interior part of the coastal region and leased out their paddy land. The Goud Saraswats, Goldsmiths who also called themselves Daivajna Brahmins, and Muslims were non-agricultural non-resident landlords. The Nadors in Ankola and the Konkan Marathas in Karwar were owner-farmers who employed agricultural labourers to carry out seasonal agricultural operations. Yennemadi, reported in 1910 that, though an inferior caste, the Nadors were an economically rising class because they were efficient cultivators of rice.[15] According to MacGregor's report, the Konkan Marathas were also a well-to-do self-cultivating landowning caste.[16]

In fact, the predominance of non-agricultural ownership of land in Uttara Kannada was admitted in the official circle also. Morarji Desai, the erstwhile Minister for Home Affairs and Revenue in the Bombay Presidency, stated that there were, in the district, 12,301 non-agricultural landlords in 1948 who owned as much as 42 per cent of the net cropped area.[17] Thus, we have clear evidence to prove our hypothesis that the dominance of non-agricultural landlords is the basic cause of tenancy in a region of paddy cultivation.

The large incidence of fixed-rent tenancy in the district was also attributable to the arduous character of agricultural operations and the reluctance on the part of the majority of non-agricultural landlords to involve themselves in these operations. The data presented in Table 2.2 show that of the various forms of tenancy, fixed-rent tenancy was predominant. The tenancy with mixed terms was also essentially fixed-rent tenancy. In 1970-71, fixed-rent tenancy (fixed cash + fixed kind + mixed terms) accounted for nearly 77 per cent of the total area under tenancy in the district, while sharecropping tenancy accounted for just 17 per cent. It can be seen from Table 2.2 itself that both in the coastal and Malnad regions the incidence of fixed-rent tenancy was very large.

Table 2.2

Proportions of Tenanted Land by Forms of Rent in Uttara Kannada, 1970-71

S. No.	*Talukas*	*Fixed Cash*	*Fixed Kind*	*Share of Produce*	*Mixed Terms*	*Others*
Coastal Region		12.7	30.8	21.3	27.5	7.7
1.	Ankola	23.7	25.5	21.5	22.8	6.5
2.	Bhatkal	6.3	51.9	7.1	22.8	11.9
3.	Honavar	11.3	14.1	33.1	34.6	6.9
4.	Karwar	10.4	41.4	13.0	19.6	15.6
5.	Kumta	10.0	30.0	24.6	33.1	2.3
Malnad Region		33.0	28.6	12.8	21.4	4.2
6.	Haliyal	11.4	24.2	44.4	17.5	2.5
7.	Mundagod	59.3	22.4	1.6	9.2	7.5
8.	Siddapur	25.8	36.8	3.1	32.2	2.1
9.	Sirsi	38.6	29.3	3.8	24.6	3.7
10.	Supa	29.7	17.1	28.3	10.2	14.7
11.	Yellapur	46.2	29.3	6.5	12.5	5.5
	District	22.6	29.7	17.2	24.5	6.0

Source: Derived from Census of Agricultural Holdings in Karnataka, 1970-71.

Dinakar Desai, who pioneered the tenancy movement in the district, made a deep study of land rents in the coastal region in the late 1930s. He found that, while in the Belgaum and Dharwad districts, the landlords involved themselves actively in agriculture by sharing costs and profits of farming, those in the coastal region of Uttara Kannada were sleeping partners because they had leased out their paddy land on fixed-rent basis.[18] Maxwell saw that the sole object of fixed-rent tenancy was to relieve the landlord from all concern in cultivation.[19] He also reported that the absence of sharecropping in many coastal villages was due to unwillingness of the non-agricultural landlords to undertake supervision over cultivation which was inconvenient and troublesome.[20] We may note here that in Haliyal taluka sharecropping tenancy, locally known as *korpalu*, accounted for over 44 per cent of the total area under tenancy in 1970-71 (see Table 2.2), because here landlords (Lingayats) were basically interested in paddy cultivation. But elsewhere in the Malnad region, the landlords were less interested in paddy cultivation so that in the paddy producing region characterised by the preponderance of non-agricultural landlords fixed-rent tenancy is predominant.

Tenancy which was deep-rooted in Uttara Kannada became a pervasive phenomenon by the turn of the 19th century. Though data to show increase in the extent of tenancy over the years are not available, we can learn from historical documents that tenancy became widespread after the advent of the British rule. Monteath noticed in 1913 that tenancy cultivation had become more common over the years in the coastal region.[21] In the up-ghat region also, there was a substantial change in the agrarian structure by the beginning of the present century leading to the growth of tenancy. In 1883, as the Kanara Gazetteer reported, in Sirsi, Siddapur and Yellapur talukas, the ownership of land was almost entirely in the hands of the Havik Brahmins.[22] In the subsequent years, however, quite a few Haviks fell to the rank of tenants.

The growth of tenancy in the district was due to two reasons: (1) usurious practices pursued by landlords, and (2) the flow of the merchant's capital into the countryside. The relatively large landlords known as *khatedars* were also village moneylenders in Uttara Kannada, and all needy husbandmen and villagers looked to them for loans.[23] The tenants used to borrow grains from their landlords and repay these grain advances after harvest. The interest charged on these advances normally amounted to 25 per cent, and, in many cases, it was as high as 50 per cent.[24] Besides Brahmin landlords, the Nadors also made gains by indulging in usury.[25]

The prominence acquired by the Goud Saraswat landlords by the middle of the 19th century as usurers due to the British policy of encouraging moneylending business was mainly responsible for the growth of tenancy. By accepting land as security, they advanced loans to illiterate peasants. In course of time, the lands of the debtors fell into the hands of the usurers, and the former became tenants on the same land which they had owned before. The Kanara Gazetteer has recorded: "Almost all large landholders still unite moneylending to their occupations. In upland Kanara, the rule is for occupants to cultivate their own lands, but everywhere there are large landholders and the process of the more ignorant cultivators being converted from occupants into tenants which is near completion in the lowland, is also in operation in upland Kanara."[26]

The penetration of the merchant's capital into the rural areas in both Malnad and coastal regions of the district also led to an increase in the incidence of tenancy. In the pre-independence period, the up-ghat region was marked by dense forests, lack of basic amenities (such as transport, communication and medical care), malarial climate and diminishing population. Indeed, in this period, population in Uttara Kannada declined more sharply than in the country as a whole, and increased only marginally when the rate of growth was much higher in the country.[27] This is seen in Table 2.3. There is clear evidence to show decline in population in the up-ghat talukas (see Table 2.4). The factors which were totally unfavourable for human habitation made the below-ghat labourers develop contempt and dislike for the rural up-ghat tract. Collins, in 1926, noticed a marked decrease in the down-ghat labour supply and the resultant rise in wages. The cost of labour had increased out of all proportions to any rise which might have taken place in the price of garden produce. So acute was the condition that many precious gardens fell into ruin and were converted into rice-fields later on.[28] Because of the increased cost of cultivation and also because of their litigiousness and lavish spending, many Haviks ran into debts. They pledged their gardens and borrowed huge sums of money from the betelnut merchants or *sawkars* in Sirsi town, mostly belonging to the Goud Saraswat community. Being unable to discharge their debt obligations, some of them sold their estates to the *sawkars*, and leased these on *mulgeni*, permanent tenancy

Table 2.3
Population Changes, 1901-1941

Year	*Population in Uttara Kannada (in lakhs)*	*Rate of Growth in the Decade (%)*	
		Uttara Kannada	*India*
1901	4.54	-	-
1911	4.31	-5.3	5.8
1921	4.02	-6.7	-0.3
1931	4.18	4.0	11.0
1941	4.41	5.6	14.2

Source: Census of India for various years.

Table 2.4
Percentage Changes in Population in Up-ghat Talukas of Uttara Kannada, 1901-1941

Year	*Haliyal*		*Mundagod*		*Siddapur*	
	Change in the Preceding Decade	*Change Since 1901*	*Change in the Preceding Decade*	*Change Since 1901*	*Change in the Preceding Decade*	*Change Since 1901*
1911	-19.0	-19.0	-14.0	-14.0	-9.3	-9.3
1921	-13.8	-30.0	-29.9	-39.7	-9.5	-17.9
1931	15.1	-19.6	20.9	-27.1	-1.6	-19.3
1941	-3.5	-22.4	-10.5	-34.8	2.2	-17.4
Year	*Sirsi*		*Supa*		*Yellapur*	
	Change in the Preceding Decade	*Change Since 1901*	*Change in the Preceding Decade*	*Change Since 1901*	*Change in the Preceding Decade*	*Change Since 1901*
1911	-10.1	-10.1	-11.4	-11.4	-10.9	-10.9
1921	-14.3	-22.9	-13.3	-23.2	-14.4	-23.8
1931	-3.4	-25.6	-5.6	-27.5	-10.0	-31.5
1941	8.3	-19.4	3.3	-25.1	-2.7	-33.3

Source: Derived from 1961 District Census Handbook, North Kanara District.

with payment of fixed cash rents. In this way, they became tenant gardeners in the early 20th century. The impact of the entry of the merchant's capital into agriculture could be gauged by the fact that by the dawn of 1920s, nearly a quarter to a third of the land in Sirsi taluka and a large area of land in adjacent villages of Mundagod and Siddapur talukas came to be owned and controlled by the merchants of Sirsi town.[29] However, the number of the tenant gardeners belonging to the Havik community was less than that of the owner-cultivators even during the 1920s. Moreover, several Haviks who became tenant gardeners owned paddy land.

In the coastal region, especially in Bhatkal taluka, the Muslims who amassed wealth through trade and commerce, found that the purchase of land was almost the only profitable mode of investing their savings.[30] By the beginning of the present century, their speculations in land extended even to the remotest forest villages.[31] Subsequently, they leased out the land mainly or even wholly, on fixed rent.

Thus, the instance of Uttara Kannada shows that in a region where non-agricultural landlords acquire prominence as usurers also, the extent of tenancy increases due to the transfer of property titles of the indebted owner-cultivators to the usurers, the former becoming the latter's tenants. It also shows that the penetration of the merchant's capital into the countryside eventually leads to the growth of parasitic absentee land-ownership. Naturally enough, the same factors which led to the spread of non-agricultural landlordism and the growth of tenancy accounted for the increasing incidence of fixed-rent tenancy in Uttara Kannada. Since the origin and growth of tenancy, particularly in the case of paddy land, were due to caste factor and usury, it was feudal in character.

A very large proportion of landlords in the district, besides being non-agricultural, were small owners. Munro vividly described how the land here was parcelled out amongst a prodigious number of proprietors who paid fixed assessment to the *Sarkar* or government.[32] As Table 2.5 shows, in 1947-48, small landowners, owning less than five acres each, represented nearly 75 per cent of the total number of landowners in the district, though they owned just 18 per cent of the land. The small pure rentiers formed nearly 70 per cent of the total number of pure rentiers, though they accounted for only 13 per cent of the total leased-out area. In Uttara Kannada, small landlords had leased-out land to small tenants.

Table 2.5
Size-Class-wise Proportions of Landowners in Uttara Kannada, 1947-48

Size of Ownership Holdings (acres)	A		B		C		D	
	No.	Area	No.	Area	No.	Area	No.	Area
Upto 5	78.6	26.1	79.9	14.5	68.8	13.0	74.4	18.1
6-15	16.0	36.3	11.4	20.7	19.2	21.4	17.4	25.8
16-25	3.5	17.7	3.5	14.6	5.3	12.4	4.2	14.5
26-100	1.8	18.8	4.9	39.2	5.8	31.4	3.6	27.3
101-500	0.1	1.1	0.3	11.0	0.9	19.5	0.4	12.8
501 and above	-	-	-	-	0.1	2.3	Neg.	1.5
Total	100	100	100	100	100	100	100	100

Notes: A = Those who cultivated land themselves with or without the help of hired labour.
B = Those who did not cultivate land themselves but supervised and directed the cultivation by hired labour.
C = Pure rentiers.

Source: Derived from the Statistical Abstract of Bombay State, 1952, Bureau of Economics and Statistics, Government of Bombay, 1954, p. 170.

Among the tenant classes in the Malnad region, numerically the most important were Halepaiks or Namdharis. Historical evidence suggests that many Namdharis, especially in Siddapur taluka, owned garden land and cultivated paddy land owned by non-agricultural Brahmin landlords as tenants.[33] The other cultivating castes, such as Karevakkals, Agers and Kotevakkals were numerically unimportant.[34] In the coastal region of the district, Halakki Vakkals, Deshbhandaris, Komarpaiks, Pandthis, Namdharis and Agers were the important cultivating castes. Majority of them made their living by hiring out labour as tenants-cum-labourers, while only a few were owners-cum-labourers.[35]

In the light of the historical evidence given above, we could identify four agrarian classes in Uttara Kannada. The Goud Saraswat and Saraswat Brahmins, Goldsmiths and Muslims and those of the Havik Brahmins who did not put in any self-employment but hired in labour to get their land cultivated, constituted the class of landlords. As already stated, there were many Haviks in the non-coastal villages of the district who, in addition to employing labourers, carried out many horticultural operations themselves; they of course hired in labour to keep their paddy land under cultivation. Such Havik gardeners, along with Lingayats, Nadors and the Konkan Marathas, formed the class of rich peasants. Halakki Vakkals, Komarpaiks, Pandthis, Deshbhandaris and Namdharis constituted numerically the largest class of poor peasants. The class of landless labourers consisted of Halakki Vakkals, Komarpaiks, etc., who did not hold any land even as tenants and the untouchables like Agers and Holers.

The overwhelming majority of the landlords and the rich peasants belonging to the Havik and Lingayat communities came from the so-called superior castes. The other rich peasants, namely, the Nadors and the Konkan Marathas belonged to the upper middle castes. The Namdharis who were poor peasants, belonged to the lower middle caste. But the vast majority of poor peasants hailed from socially backward castes. The landless labourers belonged to either backward or scheduled caste. Thus, class and caste broadly coincide in Uttara Kannada.

Regional Variations in the Extent of Tenancy

The extent of tenancy was different in different regions of Uttara Kannada. It can be seen from Table 2.6 that in 1970-71, the extent of tenancy was far more in the coastal region than in the Malnad region. The proportion of the total area under tenancy to the total operated area was more than 56 per cent in the below-ghat region, while it was not even 35 per cent in the Malnad region; it was as high as 64 per cent in Bhatkal taluka, while it was as low as 15 per cent in Supa taluka. The percentage of owned and self-operated area to the total operated area was naturally much larger in the Malnad region than in the coastal region.

Table 2.6

Extent of Tenancy in Different Regions of Uttara Kannada, 1970-71

(in % of operated holdings/area)

S. No.	*Talukas*	*No. of Operated Holdings*	*Operated Area (acres)*	*A*	*B*	*C*	*D*	*E*
1.	Ankola	11,878	29,345	24.8	27.4	75.2	72.6	51.7
2.	Bhatkal	9,280	17,149	21.4	19.7	78.6	80.3	64.1
3.	Honavar	17,089	32,741	17.1	19.0	82.9	81.0	56.2
4.	Karwar	19,466	31,399	40.5	24.8	59.5	75.2	51.8
5.	Kumta	15,425	33,499	17.4	17.1	82.6	82.9	61.4
	Coastal region	73,138	1,44,133	25.2	21.6	74.8	78.4	56.5
6.	Haliyal	5,856	49,946	61.4	54.7	38.6	45.3	30.9
7.	Mundagod	3,951	31,661	64.7	73.9	35.3	26.1	30.5
8.	Siddapur	8,848	36,894	26.3	22.4	73.7	77.6	48.9
9.	Sirsi	8,006	57,129	34.4	32.1	65.6	67.9	41.8
10.	Supa	3,240	22,402	66.9	50.3	33.1	49.7	15.0
11.	Yellapur	3,328	2,310	58.5	53.9	41.5	46.1	28.0
	Malnad region	33,229	2,00,342	46.2	47.6	53.8	52.4	34.8
	District	1,06,367	3,44,475	31.8	36.0	68.2	64.0	43.4

Notes: A = Wholly owned and self-operated holdings.
B = Wholly owned and self-operated area.
C = Partly owned and partly leased-in holdings.
D = Partly owned and partly leased-in area.
E = Total tenanted area.

Source: Derived from Census of Agricultural Holdings in Karnataka, 1970-71.

In the light of historical information, we can identify two factors that accounted for regional variations in the extent of tenancy. Before the advent of independence itself, in contrast to the large portion of the Malnad region and also the inland forest tract of the down-ghat talukas, the coastal belt was characterised by a larger and growing population. The healthy climate, in addition to the development of transport and communication, induced people to move from inland villages to the coastal tract. Early in 1913, Monteath found that in this tract, cultivation had, to all intents, reached its limit.[36] Naturally, the increasing pressure of population on land here led to a large incidence of tenancy. In the Malnad region, on the contrary, while the shortage of labour posed a serious problem, it was equally difficult to get sufficient and reliable tenants to cultivate the paddy land. The major problem of agriculture which dwarfed all others in the region was to keep the land under plough.[37] Tenancy disappeared from many rice fields simply because tenants disappeared on their own.[38]

Secondly, the incidence of tenancy was more in the coastal region because the non-agricultural landlordism was more wide-spread here than in the up-ghat region.[39] In other words, the non-cultivating non-resident landlords outnumbered the resident landlords in the coastal region, whereas, in the Malnad region, the bulk of the landowners were resident in the village in which their lands were located and belonged to the castes interested in agriculture.[40] Thus, the study of the early agrarian structure in Uttara Kannada reveals that in addition to the differences in the natural factors and density of population, the difference in the character of landlordism also account for regional variations in the extent of tenancy. The variations in the extent of tenancy was accompanied by variations in the distribution of terms and conditions in various types of tenancy in different regions of the district.

Different Types of Tenancy in Uttara Kannada

Different types of tenancy existing in Uttara Kannada were *mulgeni*, *sulgi*, *chalgeni* and *palu*. *Mulgeni* and *chalgeni* were fixed-rent tenancies, while *nadagi*, *sulgi* and *palu* were sharecropping arrangements. The terms and distribution of benefits in all these types could throw an insight into the relative bargaining strength of the two parties (i.e., the lessor and the lessee) which was, however, different in different regions of the district.

Mulgeni was a kind of permanent and hereditary tenancy. As long as the *mulgeni* tenant, popularly called Mulgenidar, paid a specified rent to the landlord, he could mortgage, sell his interest in the land, or even sublease the land.[41] *Mulgeni* contracts were always in writing and were registered. While the Kanara Gazetteer treated the Mulgenidar on par with the landlord,[42] the High Court of Bombay consistently held the view that he was only a tenant with of course superior and permanent rights in land.[43] However, there is no denying the fact that Mulgenidars were, by and large, privileged tenants enjoying strong bargaining strength *vis-a-vis* their lessors.

Historical information shows that *mulgeni* was more common in the case of the garden land than in the case of the paddy land.[44] As M.J. Desai wrote in 1935: "The *mulgeni* lease is, therefore, very well suited for garden cultivation, where this year's yield represents the fruits of the labours of the past and where the fruits of this year's labour are reaped in the next 40 or 50 years."[45] Collins also found that *mulgeni* was particularly suitable for garden cultivation in which the tenant required to sink a large amount of capital was not willing to lease in land for a short period and the landlord after spending his capital on the garden wanted a fixed return.[46] Thus, the instance of *mulgeni* showed that permanent lease contract is suitable for those of the commercial crops the cultivation of which requires constant care and heavy long-term investment by the tenant.

The rent paid by the Mulgenidar was determined by the share of the each party in the investment in the land, for gardens in bad state were let out on low rents, while those in good state were leased out on high rents.[47] Besides, *mulgeni* rent reflected the expectations of the two parties at the time of the execution of the lease. As Mathrani writes: "Thus, in the period of high prices the rents agreed upon were very high and a large number of leases had to be resold or mortgaged in the years of slump in order to pay rents."[48] Further, he found that *mulgeni* agreements were normally terminable at the will of the tenant, unless one of the conditions of the contract was that it could only be terminated by the consent of both the parties.[49] The overwhelming majority of the Mulgenidars cultivating the garden land paid rent in terms of cash. This shows that cash rents were predominant in the case of commercial crops.

There were some ancient *mulgeni* agreements that stipulated fixed cash rents which were either equal to or only slightly in excess of the land revenue. In later *mulgeni* deeds, executed especially after the introduction of survey operations, the rents were higher because these deeds contained a stipulation that if the assessment or revenue were raised, the Mulgenidar must pay the enhanced assessment. However, in the normal course of things, this condition did not pose any serious difficulty for the Mulgenidar, as the land assessment was liable to revision at intervals of thirty years only.[50]

The description given above relates to normal or genuine *mulgeni* leases. However, the distribution of terms and conditions in *mulgeni* leases which were the direct outcome of usury reflected the weak bargaining strength of the tenant. Such leases were interest-leases, for the interest that the usurer wanted to exact from the debtor was disguised as rent.[51] While rents in normal *mulgeni* leases varied between Rs. 5 and 15 per acre, interest rents varied from Rs. 40 to 60.[52] To keep the debtor-cum-tenant always in his grip, the creditor-cum-landlord imposed a condition that the tenancy could be terminated only with his consent. Thus, the bargaining capacity of the tenant in the interest-lease was very weak when compared to that of the tenant in the genuine *mulgeni* lease. Yet, the security of tenure and fixity of rent were the privileges enjoyed by all Mulgenidars.

A few *mulgeni* leases, the origin of which was lost in antiquity provided that in addition to cash rent, the tenant must contribute small but fixed quantities of pepper, betelnut and cardamom as a token of honour and respect for the landlord. There were also some *mulgeni* agreements with an additional stipulation on the tenant's part to give the landlord a small part of his labour without payment for one or two days in a year.[53] We venture to state that such additional stipulations were feudal in character. Thus, we have evidence to show that in a traditional land-based social order, the possibility of feudal elements characterising the contractual terms cannot be ruled out. However, the tenant in the *mulgeni* leases with feudal elements was required to pay only nominal cash rent.[54] Therefore, these leases had no sharp exploitative edge.

Though *mulgeni* was the predominant form of lease in the case of garden land, a few *chalgeni* leases also existed in respect of gardens in the Malnad region. But in most cases, *chalgeni* was a sort of feeler, the tenant trying to find out what the net profits were like so that he could, after a few months experience, decide on the terms of *mulgeni* lease for the same land.[55] There were, however, some cases of regular *chalgeni* leases, where the garden land had been let year after year to the same tenant, but where the landlord was reluctant to part with full proprietory rights by leasing land on *mulgeni*. In addition to economic factors, the caste factor also influenced the distribution of terms and benefits between the two parties in these cases. While the Havik landlords leased out gardens to tenants belonging to their own caste for a nominal cash rent, they demanded artificially high rents from the non-Havik tenants like Namdharis, Karevakkals and Kotevakkals. Mathrani specifically states that the bargaining strength of the tenant, in these cases, depended to a large extent on the caste factor.[56] If the two parties belonged to the same caste, the tenant secured equitable terms; on the contrary, if the landlord belonging to the so-called superior caste leased out land to a tenant belonging to the so-called inferior caste, the contractual terms were unfavourable to the latter. However, the Havik landlord was duly penalised for his rapacity of demanding high rent from the non-Havik *chalgeni* tenant, for when the latter found that the rents became too high, he reaped the crops without carrying out any of the usual operations necessary for the future prosperity of the garden and left the leasehold. In 1935, M.J. Desai met with several cases where the *chalgeni* tenant having nothing at stake had disappeared and the landlord was left with a much deteriorated garden on his hands.[57] This indeed showed that *chalgeni* was not a suitable form of lease for the garden land.

Nadagi and *sulgi*, tenancies of quasi-permanent nature, existed on a few coconut plantations in the coastal area of Kumta and Honavar talukas which did not require intensive and continuous care necessary in the case of spice gardens in the Malnad region.[58] The rent payable by the *nadagi* tenant, popularly called Nadagidar, was fixed generally at one half of the produce, and in a few cases, in cash. The landlord had to bear the expenses of planting the trees, and the Nadagidar, the expense of rearing them. When full-grown trees were made over to a tenant, the tenancy was called *sulgi*, and in this case, the tenant received one-third of the produce. Both *nadagi* and *sulgi* could be terminated at the will of the tenant but he could not be ousted by the landlord, unless it was proved that the property had suffered from neglect. As the Kanara Gazetteer reported: "These tenures appear to be declining as they give rise to numerous disputes."[59] However, the terms and distribution of benefits in both *nadagi* and *sulgi* indicated that greater the share of the landlord in investment and maintenance of the leased land, greater was the rent and *vice versa*. Therefore, the landlords in these forms of lease were not parasitical elements.

Mulgeni, *nadagi* and *sulgi* were not predominant forms of lease in Uttara Kannada obviously because it is primarily a rice-producing region. The most common form of lease in the case of paddy land which also accounted for very high proportions of tenancy in the lower size-classes was *chalgeni* or tenancy-at-will.[60] Indeed, the instance of *chalgeni* showed that short-term or temporary lease contracts are predominant in the case of seasonal crops like paddy if its cultivation does not necessitate any significant long-term investment. A Chalgenidar might hold land under the original proprietor or Mulgenidar. The rights of a Chalgenidar possessing a written lease were noted in the Records of Rights. But there were also oral *chalgeni* agreements which obviously remained unrecorded.[61] Historical documents show that the overwhelming majority of the *chalgeni* tenants holding the paddy land were bound by contract to pay fixed produce rent. The Kanara Gazetteer stated: "In Supa, the rent is ordinarily fixed in cash; elsewhere, except for garden land, it is almost always paid in grain."[62] Even Mulgenidar cultivating paddy land, though of course very few in number, were required to pay grain rent.[63]

The produce-rent paid by the *chalgeni* tenant was usually determined according to the sowing capacity or the see-scale which reflected the fertility of the leased land. Buchanan reported that the tenant paid, for every *moray* of seed sown, two *morays* of rice for land of the best quality, 1.5 *morays* for average or 'middling' land and one *moray* of rice for the worst or the *makki* land.[64] Even in the early 1920s, the kind rent payable by the Chalgenidar depended mainly on the fertility of the soil.[65] In fact, there is clear historical evidence to show that the greater the fertility of the soil, the greater the rent and *vice versa*. However, there were striking variations in the distribution of terms and benefits in *chalgeni* and *palu* (sharecropping) in different regions of the district which naturally indicated regional variations in the bargaining strength of the tenant cultivating the paddy land. In the Malnad region, excluding the rice tract of Siddapur taluka, the tenant had strong bargaining power. As Mathrani wrote in 1942: "In fact, labourers and tenants are the masters of the situation in Yellapur taluka."[66] Collins' illuminating account of the agrarian relations also lends support to the fact that in many villages of Sirsi, Mundagod and the garden tract of Siddapur talukas, the tenant could virtually dictate his own terms because the farm wages were high.[67] The normal produce rent varied between 25 and 33 per cent of the gross yield of the paddy land which was by no means excessive when compared with the rent paid by the Chalgenidar in the coastal tract of the district.[68] In a few pockets of Mundagod and Yellapur and in many villages of Haliyal, a system of sharecropping was prevalent where the landlord ordinarily paid the assessment, 50 per cent of the cost of seeds and the entire cost of bullocks and manure; while the grain was equally divided between the two parties, the landlord retained the entire straw. In the course of his enquiry, Collins found that the crop-sharing arrangement was, in reality, a losing game for the landlord in the Malnad region. But it was necessary because of the difficulties associated with the cultivation of paddy land. As Collins observed: "...I am of the opinion that it is a suitable form of lease which has stood the test of time. It exists side by side in some villages with the ordinary system of rents in kind and the profit of the landlord approximates to the same level in both cases."[69] This clearly proves our hypothesis that if sharecropping and fixed-rent tenancies exist side by side in the same region in respect of the same crop, the gains shared by tenants under both the systems tend to be more or less the same, though terms apparently differ. In fact, in the large part of the Malnad region of Uttara Kannada, both the Chalgenidar and the *palu* tenant secured equitable contractual terms.

The rice tract of Siddapur taluka, where the climate was healthy, presented a picture contrary to the rest of the up-ghat region. Collins divided Siddapur taluka into two distinct parts: (1) rice tract covering an area of 83.7 sq. miles, spread over 45 villages, with a density of population of 155 per sq. mile, and (2) garden tract covering an area of 248.2 sq. miles spread over as many as 157 villages, with a density of population of only 77 per sq. mile.[70] Mathrani noticed that agricultural wages in the rice tract were comparatively low, and keen competition among poor peasants, namely, Namdharis and Kotevakkals to lease in land resulted in high farm rent. Here, the *chalgeni* tenant was at the mercy of the landlord, who gave him an advance of grain and cash, and owing to the high rent, often showed a debit balance in his account. He quite often took away the entire crop and adjusted its cash value to the amount due from the tenant. [71] Indeed, in the rice tract of Siddapur, the landlords exploited the poor peasants through two means, viz., rack-renting and usury. Here, the contradictions between the landlord and the tenant were very sharp.

Thus, in the large part of the Malnad region where the extent of tenancy was less, the tenant enjoyed strong bargaining strength *vis-a-vis* the landlord. This strength depended, among others, on agricultural wages.

In the coastal belt of the below-ghat talukas, the land tenancy system exhibited its classic drawbacks in the latter half of the 19th century itself. Here the exploitative edge of the *chalgeni* system widely prevalent on the paddy land was very sharp. As the Kanara Gazetteer observed: "In Kumta, Honavar and Bhatkal, the proprietors usually let the land from year to year and are hard and exacting landlords taking from the yearly tenants at least as much as the whole of the produce."[72] In the coastal villages of Karwar and Ankola also, the tenants were exploited as they had to part with not less than 80 per cent of the produce of paddy land as rent.[73] Here, even Mulgenidars cultivating paddy land were also required to pay exorbitant rents.[74] There was clear official admission about rack-renting in the coastal tract. Swifte, the Collector of Kanara, observed in 1908: "While admitting the high letting value of the lands in villages on or near the coast, I am inclined to think that in not a few cases it is unnaturally high, and more than the land is really entitled."[75] Dinakar Desai came to the conclusion that the tenants, except those cultivating the *gazni* land, were required to bear the entire cost of cultivation which was never less than 50 per cent of the gross produce. He found that the bulk of the *chalgeni* tenants were left with almost nothing for all the toils and troubles associated with paddy cultivation.[76]

The bargaining strength of the tenant in the coastal village was weak because the farm wages were low. In the inland part of the below-ghat talukas, where the density of population was less and the agricultural wages were comparatively high the contractual terms for the Chalgenidar were much better. Maxwell calculated that the average rent in terms of money was Rs. 21 per acre in the coastal villages of Kumta taluka whereas in the inland villages it was as low as Rs. 8 or, in some cases, even Rs. 5.[77]

Only in very few villages *palu* was prevalent. The various systems of *palu* in vogue respectively involved payment by tenant of one-half, five-ninth, three-fourth and two-third of the outturn of paddy, after deducting seeds, the only input supplied by the landlord.[78] The first variety was found mainly in the interior villages, while the second and fourth varieties were very rare.[79] The third variety was common in the coastal villages.[80] However, these evidences are sufficient to show that the half-share system cannot be taken as a universally existing practice. Another important fact that in those villages where *chalgeni* with fixed produce rent and *palu* co-existed, the gains shared by tenants involved in both these systems were almost equal. As Maxwell wrote: "...it is inconceivable that both systems should survive if one left the tenant in a much better position than the other."[81] Thus, the historical documents throwing light on the structure of agrarian relations in the coastal region also lend support to our hypothesis that if fixed-rent and crop-sharing exist side by side in the same region in respect of the same crop, the gains shared by tenants under both the systems tend to be more or less the same, though terms apparently differ. In the entire coastal belt of Uttara Kannada, the position of both the Chalgenidar and the *palu* tenant was highly vulnerable.

Yennemadi's observation on the plight of tenants in Ankola taluka indeed indicated the conditions which prevailed in the entire coastal belt of the down-ghat talukas of Uttara Kannada. To quote: "The letting value is high. This is to a considerable extent due to rack-renting facilitated by the fact that the villages are overcrowded with cultivating classes, that they are disinclined to move further to the inland villages although they can get land for cultivation on more favourable terms, and that the love of villages which they inhabited from time immemorial and the deep-rooted prejudice against the inland villages incline them to be content with whatever share they may get of the produce of land they cultivate."[82] He also noticed the glaring contradictions between the landlord and the tenant in the region and predicted in 1910 itself that rack-renting would be, sooner or later, resisted, and his prediction came true after 30 years.

Gazni Land

Indeed, the cultivation of the *gazni* or *khar* land which constituted 6 per cent of the total area under paddy in the district in 1950 was, in the coastal region, a challenging task. Located on the mouths of rivers and small creeks, these pieces of land were exposed to flood from sea-water, particularly at the time of high tides. The traditional technique employed in preventing the entry of saline water was to raise heavy earthen dykes or embankment with sluice gates constructed into them. The sluice gates were opened in order to facilitate the entry of the flood water of the rivers into the paddy fields during the rainy season which resulted in accumulation of silt with a high content of humus and other soil nutrients. But these sluice gates were closed to prevent the entry of saline water. The bunds and sluice gates were renovated every year. But there was the danger of a serious breach occurring in the embankments, thereby impairing cultivation of land for a period of four to five years. As MacGregor wrote in 1908: "If the season is so propitious that sufficient flood water charged with silt is available, the embankment has stood and there has been no intrusion of salt water very good crops can be raised, but there are many factors necessary to make the perfect season."[83] There were in all 14 major *gaznis*, each *gazni* being named after the village where it was located.

The ecological conditions surrounding the *gaznis* were such that heavy investment and reclamation measures were necessary every year which the tenants could not afford. Salt water was a common problem incapable of being solved by individual efforts. Therefore, the *gazni* land had a very organised joint farming system since centuries, though there were no registered cooperative societies. To organise and supervise all the operations relating to the construction and repair of embankments and sluice gates, there used to be a chief executive for each *gazni* known as *patgar*. He used to ensure the contribution of labour by tenants and finance by landlords needed to meet the expenses of materials necessary for embankments and sluice gates.[84]

No doubt individual landlords had their well demarcated lands but the holding of each was described not in acres but in percentage of total area of the *gazni* concerned. In spite of the varied productivity of different plots, the total production in the *gazni* used to be pooled together and divided according to the share of a particular individual in the total *gazni* land and he used to get his due share whether his particular plot had fared well or not. Thus, there was common pooling of resources and equitable distribution of product.

The total produce in the *gazni* land used to be brought to the common threshing yard locally known as *kala*. At the time of threshing, all those who had an interest in the *gazni*, whether as tenant or as landlord, used to be present. After threshing, a certain proportion of paddy used to be set apart for unforeseen calamities, and the remainder used to be distributed amongst the landlords and tenants equally. The *gazni* land was mostly owned by non-agricultural landlords residing in Karwar, Kumta, Ankola, Honavar and Bhatkal towns. They found the investment in the *gazni* land highly profitable because they exported rice to Goa where it could be sold at a very high price.[85] The instance of *gazni* land, thus, shows that in extreme cases, even non-agricultural absentee landlords evince interest in agriculture by way of investment, if they are sure of getting adequate return on it.

Development of Capitalism in Plantation Sector

There is no dispute in the fact that there was development of capitalism in horticulture or the plantation sector in Uttara Kannada. All the three characteristic features of agrarian capitalism, viz., employment of wage labour on a large scale, production of marketable surplus and its realisation through sale were present in this sector. The fact that *mulgeni* was predominant in the case of gardens shows that tenancy can be a vehicle for the development of agrarian capitalism. Furthermore, the form as well as terms and conditions of tenancy in the villages of Sirsi, Siddapur and Yellapur talukas changed with the changing agrarian structure. The Havik population in the district decreased from 40,560 in 1901 to 36,109 in 1921 or by 11 per cent. Most of the decrease was contributed by the above compared with the below-ghat portion of Uttara Kannada.[86] This decline, along with the acute difficulties of cultivation already mentioned, resulted in an increase in the area of gardens held by non-Haviks, who, because of their long association with the Haviks, learnt the art of garden cultivation. Historical information shows that many Namdharis and Kotevakkals who belonged to the inferior castes and even untouchables like the Holers, who were just labourers to begin with on the gardens owned by the Haviks, secured *chalgeni* rights on the same lands later. Subsequently, they could secure *mulgeni* and even proprietory rights and became capitalist farmers like the Haviks.[87] Collins noted that some Haviks who were *chalgeni* tenants on gardens in the past invested their capital after securing *mulgeni* rights in the early twentieth century.[88] Some *mulgeni* leases were marked by the presence of feudal elements; yet, they were instrumental in the development of capitalism. For instance, in some villages of Yellapur taluka almost all garden lands owned by temples were given out on *mulgeni* leases for a nominal cash rent and some nominal conditions of free service necessary for carrying out 'puja' (worship) operations.[89] This shows that tenancy, even if marked by the presence of feudal elements, can facilitate the development of agrarian capitalism if its terms and conditions provide incentive for investment in agriculture. Thus, the experience of the early agrarian structure at least in up-ghat Uttara Kannada reveals that tenancy is not a purely feudal phenomenon; on the contrary, it is quite compatible with the capitalist mode of production.

The foregoing account of the early agrarian structure reveals that both capitalist and feudal modes of production existed in Uttara Kannada. The plantation sector was characterised by capitalist development and predominance of *mulgeni*. *Mulgeni* system was, of course, not completely free from defects. However, the drawbacks of this system manifested themselves only when there was a severe fall in the prices of garden produce and consequently, garden cultivation ceased to be profitable. The Gazetteer for Uttara Kannada (1985) clearly shows that with the steady increase in the prices of garden crops in the post-independence period, the picture for gardeners became rosy;[90] and there is every reason to think *mulgeni* emerged as an instrument of development of horticulture in this period. Besides, the perfect security of tenure and fixity of rent were the singular advantages available to the *mulgeni* tenant. In contrast, the tenant in *chalgeni* system which was predominant in the case of paddy, had neither of these advantageous. This was the reason as to why many rack-rented *chalgeni* tenants cultivating paddy land in the overcrowded coastal villages of Karwar taluka were eager to take out *mulgeni* lease, while their landlords were unwilling to lose the chance of increasing the rent on some future occasion by conceding the formers' desire.[91] However, even *mulgeni* tenants cultivating paddy land in the purely coastal villages were also required to pay excessive rent, though it was fixed for good and they were far outnumbered by *chalgeni* tenants whose position was still worse. The mode of production in the case of paddy was essentially feudal, and the tenancy system fully exhibited its exploitative edge in the coastal belt and also in the rice tract of Siddapur taluka. The landlords owning land in these parts of the district were parasitical elements through and through. The bulk of them belonged to so-called superior castes who had leased out their land amongst a large number of poor peasants belonging to so-called inferior castes and exploited them through rack-renting and usury. In other words, in these parts of Uttara Kannada, the contradictions in the agrarian structure were acute which led to the outbreak of tenancy movement in 1940. The movement was essentially a struggle of the poor peasants against the exploitative feudal landlords.

To sum up, high level of tenancy in Uttara Kannada was the outcome of the predominance of non-agricultural non-resident landlords. The greater proportion of lessors were small landowners though, of course, they accounted for a very small proportion of leased land. The tenants and small owner-cultivators were subject to the usurious control of landlords and rich peasants, namely, Nadors. There was a complex pattern of tenancy. *Chalgeni* with fixed kind rent was predominant in the case of the paddy land. There were marked regional variations in the extent and terms of tenancy. The incidence of tenancy was far more in the coastal region than in the up-ghat area. In the large part of the Malnad region and the inland villages of the down-ghat talukas, the farm wages were high, while the rent paid by both the Chalgenidar and the *palu* tenant was low. On the contrary, in the coastal tract and the rice tract of Siddapur taluka, farm wages were low, while the land rent was very high. Indeed, in these parts of the district, the land tenancy system exhibited glaring contradictions.

Notes

1. In the judgement delivered in Vaikuntha Bapuji Vs. Government of Bombay, the Bombay High Court makes a special mention of the system of private property in land in the North Kanara district. See *The Bombay High Court Report*, Vol. 12, 1875, pp. 52, 132, 136, 153, 156 and 159.
2. *Ibid.*, p. 132.
3. *The Gazetteer of the Bombay Presidency*, Vol. 15, Part II, Government Central Press, Bombay, 1883, p. 185.
4. *The Bombay High Court Report*, *op.cit.*, p. 78.
5. Francis Buchanan, *A Journey from Madras through the Countries of Mysore, Kanara and Malabar*, Vol. 2, Higginbotham and Co., Madras, 1870 (Second Edition), p. 298.
6. *The Land Assessment and Landed Tenures of Kanara: Abstracts of Correspondence between the Board of Revenue and the Collectors of Kanara*, German Mission Press, Mangalore, 1853, p. 71.
7. Quoted in Narendar Pani, *Reforms to Pre-empt Change: Land Legislation in Karnataka*, Indian Institute of Management, Bangalore, 1981, (mimeo), p. 38.
8. The Gazetteer of the Bombay Presidency, *op.cit.*, p.4.
9. Ravenshaw, who succeeded Munro in 1801, was moved by seeing the "extra-ordinary labour" necessary for cultivating lands in Kanara. See The Land Assessment and Landed Tenures of Kanara, *op.cit.*, p. 85.
10. The Gazetteer of the Bombay Presidency, *op.cit.*, p.2.
11. K.S. Shivanna, *The Agrarian System of Karnataka (1336-1761)*, The Director, Prasaranga, Mysore University, 1983, pp. 107-108.
12. G.F.S. Collins, *Introductory Report to the Re-Settlement Reports of the Sirsi, Siddapur and Yellapur Talukas and the Mundagod Mahal of the Kanara Collectorate*, Government Central Press, Bombay, 1926, p. 19.
13. The Gazetteer of the Bombay Presidency, *op.cit.*, p. 3.
14. *Ibid.*, pp. 2-4.
15. V.D. Yennemadi, *Papers relating to the Second Revision*

Settlement of the Ankola Taluka of the Kanara Collectorate, Government Central Press, Bombay, 1917, p. 9.

16. G.L. MacGregor, *Papers relating to the Second Revision Settlement of the Karwar Taluka of the Kanara Collectorate*, Government Central Press, Bombay, 1913, p. 4.
17. Bombay Legislative Assembly, *Debates: Official Report*, Vol. 12, Part 2, February 18, 1948, Government Central Press, Bombay, 1948, p. 23.
18. Dinakar Desai, *Land Rents in Kanara*, Servants of India Society, Bombay, 1940, pp. 3-6.
19. R.M. Maxwell, *Papers relating to the First Revision Settlement of the Kumta Taluka of the Kanara Collectorate*, Government Central Press, Bombay, 1918, p. 24.
20. R.M. Maxwell, *Papers relating to the First Revision Settlement of the Honavar Taluka and Bhatkal Petha of the Kanara Collectorate*, Government Central Press, Bombay, 1921, pp. 13-14.
21. G. Monteath, the Collector of Kanara, quoted in R.M. Maxwell, 1921, *op.cit.*, p. 91.
22. The Gazetteer of the Bombay Presidency, *op.cit.*, p. 2.
23. *Ibid.*, p. 28.
24. *Ibid.*, p. 29.
25. *Ibid.*, p. 18.
26. *Ibid.*, p. 186.
27. M.V. Nadkarni, Syed Ajmal Pasha and L.S. Prabhakar, *Political Economy of Forest Use and Management in the Context of Integration a Forest Region into the Larger Economy*, Institute for Social and Economic Change, Bangalore, 1987 (mimeo), p. 50.
28. G.F.S. Collins, *op.cit.*, p. 41.
29. *Ibid.*, pp. 17 and 56.
30. The Muslims scrupled to indulge in usury on religious grounds. See the Gazetteer of the Bombay Presidency, *op.cit.*, p. 2.
31. R.M. Maxwell, 1921, *op.cit.*, p. 8.
32. The Land Assessment and Landed Tenures of Kanara, *op.cit.*, p. 21.
33. K.P. Mathrani, *Papers relating to the Second Revision Settlement*

of the Siddapur Taluka of the Kanara Collectorate, Government Central Press, Bombay, 1942, pp.6-7.

34. G.F.S. Collins, *op.cit.*, p. 19.
35. The Gazetteer of the Bombay Presidency, *op.cit.*, pp. 2-4.
36. Quoted in R.M. Maxwell, 1921, *op.cit.*, p. 94.
37. G.F.S. Collins, *op.cit.*, p. 10.
38. *Ibid.*, p. 14.
39. *Ibid.*, p. 17.
40. Hegde found that in the below-ghat talukas, non-agricultural absentee landlordism was extensive, while in the up-ghat talukas a very large number of landowners were residents in the villages where their lands were located. See S.S. Hegde, *A Critical Study of the Working of Land Tribunals in Uttara Kannada District*, LL. M. Thesis submitted to the Karnataka University, Dharwad, 1979 (unpublished), pp. 81-82.
41. The Land Assessment and Landed Tenures of Kanara, *op.cit.*, p. 71.
42. The Gazetteer of the Bombay Presidency, *op.cit.*, p. 186.
43. In Vaikuntha Raya Vs. Shivaram Bhat (1883), Narayan Vs. Shri. Ramachandra Deva (1903), Narayan Vs. Nagappa (1910) and Ganapathi Vs. Naghabhatta (1917), the High Court of Bombay treated Mulgenidars as permanent tenants and not as landlords, or even as sub-lords. See The *Bombay Law Reporter*, Vol. 53. 1950, pp. 315-318.
44. Mathrani came across just one Mulgeni lease in the whole of Mundagod taluka, which is, even now, a rice granary of Uttara Kannada. See K.P. Mathrani, *Papers relating to the Second Revision Settlement of the Mundagod Mahal of the Kanara Collectorate*, Government Central Press, Bombay, 1942, p. 17.
45. M.J. Desai, *Papers relating to the Enquiry into the Current Revision Settlements (Garden Lands) of Siddapur, Sirsi, Yellapur Talukas including Mundagod Mahal of the Kanara Collectorate*, Government Central Press, Bombay, 1937, p. 13.
46. G.F.S. Collins, *op.cit.*, p. 52.
47. K.P. Mathrani, *Papers relating to the Second Revision Settlement of Sirsi Taluka of the Kanara Collectorate*, Government Central

Press, Bombay, 1942, p. 27.

48. *Ibid.*
49. *Ibid.*
50. The Gazetteer of the Bombay Presidency, *op.cit.*, pp. 186-187.
51. R.M. Maxwell, 1918, *op.cit.*, p. 18.
52. M.J. Desai, *op.cit.*, p. 57.
53. G.F.S. Collins, *op.cit.*, p. 53.
54. K.P. Mathrani, *Papers relating to the Second Revision Settlement of the Yellapur Taluka of the Kanara Collectorate*, Government Central Press, Bombay, 1942, p. 21.
55. M.J. Desai, *op.cit.*, p. 13.
56. K.P. Mathrani, The Report on Sirsi Taluka, *op.cit.*, p. 25.
57. M.J. Desai, *op.cit.*, p. 13.
58. The Gazetteer of the Bombay Presidency, *op.cit.*, p. 186.
59. *Ibid.*
60. *Ibid.*, pp. 186-187.
61. *Ibid.*, p. 188.
62. *Ibid.*
63. See, for instance, R.M. Maxwell, 1918, *op.cit.*, p. 18 and also G.L. MacGregor, *op.cit.*, p. 4.
64. 1 Moray = 1.224 bushels, See Francis Buchanan, *op.cit.*, p. 299.
65. G.F.S. Collins, *op.cit.*, p. 49.
66. K.P. Mathrani, The Report on Yellapur Taluka, *op.cit.*, p. 12.
67. G.F.S. Collins, *op.cit.*, pp. 58-60.
68. *Ibid.*, p. 58.
69. *Ibid.*
70. G.F.S. Collins, *Revision Settlement Report of Siddapur Taluka of the Kanara Collectorate*, Government Central Press, Bombay, 1926, p. 3.
71. K.P. Mathrani, The Report on Siddapur Taluka, *op.cit.*, p. 23.
72. The Gazetteer of the Bombay Presidency *op.cit.*, p. 12.
73. V.D. Yennemadi, *op.cit.*, p. 10 and G.L. MacGregor, *op.cit.*, p. 12.

74. G.L. MacGregor, *op.cit.*, p. 10.
75. L.C. Swifte, quoted in MacGregor, *op.cit.*, p. 37.
76. Dinakar Desai, *op.cit.*, p. 8.
77. R.M. Maxwell, 1918, *op.cit.*, pp. 45-48.
78. *Ibid.*, pp. 22.
79. *Ibid.*
80. *Ibid.*
81. *Ibid.*
82. V.D. Yennemadi, *op.cit.*, p. 20.
83. G.L. MacGregor, *op.cit.*, p. 6.
84. N. Rajagopal Rao, "Land Reclamation in North Kanara", in N.R. Rao (ed.), *Land Utilisation and Agriculture in North Kanara*, Planning Forum, Kanara College, Kumta, 1965, p. 62.
85. The rice prepared from the coarse red variety of paddy called "Kagga" grown on the *gazni* land was very popular in Goa, G.L. MacGregor, *op.cit.*, p. 6.
86. Apart from the malarious nature of the tract and the unhealthy situation of their houses, the caste customs like early marriage of women eventually leading to their widowhood contributed to the decrease in the Havik population. See G.F.S. Collins, Introductory Report, *op.cit.*, pp. 41-42.
87. *Ibid.*, p. 56.
88. *Ibid.*
89. K.P. Mathrani, The Report on Yellapur Taluka, *op.cit.*, p. 21.
90. Suryanath Kamath (ed.), *The Gazetteer for Uttara Kannada District*, Government of Karnataka Publication, Bangalore, 1985, p. 562.
91. G.L. MacGregor, *op.cit.*, p. 4.

3

Tenancy Movement and Legislations in the Pre-Reorganisation Period

Tenancy movement, with a varying degree of intensity, characterised the agrarian scene of Uttara Kannada for over a period of thirty years. Besides raising the bargaining strength of the tenant *vis-a-vis* the landlord and bringing to the fore the urgent need for reform, the movement brought about a sharp rise in the consciousness of a large number of poor peasants in the district. Indeed, the experience in Uttara Kannada also shows that only a comprehensive class conscious organisation of the toiling tenantry and its militant intervention is the prerequisite for effective implementation of tenancy reform measures.

The impact of the peasant movements and legislations in Uttara Kannada could be analysed in two distinct phases, namely, the pre-reorganisation period and the post-reorganisation period. This chapter deals with the pre-reorganisation period (period covering the reorganisation of states in November 1956), while the next chapter is concerned with the period that followed reorganisation.

Background of Tenancy Movement

Prior to the implementation of the Bombay Tenancy Act of 1939, the relations between landlord and tenant in the Bombay Province of which Uttara Kannada (then called North Kanara) was a part, were governed by Section 83 of the Bombay Land Revenue Code of 1879 and the provisions of this section, in essence, were such that they merely gave legal recognition to customary tenancy practices with all their attendant inequities and economic evils. There was no restriction on the power of the landlord to enhance rent and there was no security for the tenant.[1]

The Congress Ministry, as a popular measure, moved a bill called the Bombay Tenancy Bill which was accepted by the legislature in September 1939. It introduced a new concept, *viz.*, 'protected tenant'.[2] A protected tenant was defined as one who held the land continuously for a period of not less than six years immediately preceding 1 January, 1938. He could not be ousted unless he committed offences such as non-payment of rent and sub-letting of land. The unprotected tenant was also to get some security, for the bill provided that no lease could be for less than ten years. Rent was limited to the agreed rent or rent payable according to usage or to 'reasonable' rent as determined by the Mamlatdar (now called Tahasildar) in case of dispute between the landlord and the tenant. Though the Bill did not fix the maximum rent payable by the tenant, it provided that the Provincial Government could, from time to time, fix it in any particular area.

The bill contained a provision that the landlord could resume the land from the protected tenant for personal cultivation. Personal cultivation was so loosely defined that even an absentee landlord could resume the land and personally cultivate it just by keeping a salaried servant or manager without involving himself in agriculture in any manner.[3] This was indeed a loophole in the bill. It also made a provision for voluntary surrender of land by the tenant to the landlord which was bound to adversely affect the former who could be pressurised by the latter to surrender the leasehold 'voluntarily'. Thus, the bill was a half-hearted and, in fact, landlord-oriented measure.

Before the bill received the assent of the Governor-General of India and became a law, the Congress Ministry resigned in October 1939. Though there was a provision that the Act would come into force in any area on such date as the Provincial Government might decide, the Congress Ministry had the intention of applying it throughout the Bombay Presidency at the same time.[4] But, with the change in the political circumstances, the news spread that there was no likelihood of the Act being put into effect everywhere in the Bombay Presidency simultaneously. To relieve the poor agriculturists from the shackles of debt, the Congress Ministry had piloted a measure called the Bombay Agricultural Debtors' Relief Bill of 1939 which was also to receive the assent of the Governor-General. Its provision extending relief to the tenants who were in arrears of rent assumed much significance as far as Uttara Kannada was concerned because, here many tenants were facing rent recovery suits filed by their exacting landlords. However, after the resignation of the Congress Ministry, the news spread that the implementation of the Agricultural Debtors' Relief Act was also likely to be piecemeal. The urgent need for legislative measures to improve the lot of the poor peasants in Uttara Kannada became manifest with the outbreak of the tenancy movement.

The Origin and Development of Tenancy Movement

With the outbreak of the tenancy movement in Ankola in March 1940, Yennemadi's prediction made as early as 1910 came true. In the light of his study of agrarian relations in Ankola taluka, he had observed that rack-renting would be, sooner or later, resisted.[5] Therefore, it was no accident that Ankola became the birthplace of the tenancy movement in Uttara Kannada. The initial stimulus for the movement, however, came from an external source, for Dinakar Desai, a native of Ankola and a life-member of the Bombay branch of the Servants of India Society, was inspired by his mentor N.M. Joshi, one of the pioneers of the Indian trade Union movement, to set the tenants against landlords.[6] When Desai gave a call to the tenants in Ankola to resist rack-renting collectively, they readily responded to it as if it was the very thing for which they were anxiously waiting since long. Thus, far from being an accidental or sporadic outburst, the movement was the logical and inevitable outcome of the exploitative land tenancy system in the region. There was a large meeting of the poor peasants like the Halakki Vakkals, Agers, Komarpaiks and Namdharis on March 16, 1940 which culminated in the establishment of the Ankola Taluka Ryot Sangh (ATRS).[7] Desai informed the peasants that the Tenancy Bill which was going to become an Act shortly was not likely to be applied to Uttara Kannada without their sustained struggle. He also told them in a convincing manner that they had to ceaselessly fight for the execution of the Agricultural Debtors' Relief Act in the district.[8]

The Tenancy Bill received the assent of the Governor-General and became a law on 2 April, 1940.[9] But the Provincial Government now led by the Governor did not specify the date on which the Act would be put into effect. Nor did it announce whether the Act would be applied to the whole province at the same time. The ATRS, making a strong case for immediate execution of the Act in Uttara Kannada, demanded that the maximum rate of rent here should be restricted by government notification to one-third of the gross yield of the paddy land. It also urged the immediate execution of the Agricultural Debtors' Relief Act in the district. Processions and meetings were held in many villages in which *chalgeni* and *palu* tenants raised red flags and sang songs composed by Desai which decried the evils of feudal landlordism. *Mulgeni* tenants cultivating paddy land also joined them because they too had to pay excessive rents in kind. However, their number was less because *mulgeni* was not a predominant form of lease in the case of paddy land.

The ATRS successfully prevailed upon its members to refuse to pay more than one-third of the yield as rent. The tenants in the whole village of Ambarkodla refused to pay rent when their landlords took an uncompromising stand by demanding the pre-determined rent. The landlords who were defied for the first time, saw in Desai the root cause of their trouble. Calling him *havali manja* (a terribly dangerous man), they alleged that the ATRS under his leadership was sowing the seeds of class struggle.[10]

The district level conference held at Ankola on May 12, 1940, which was attended by nearly five thousand peasants gave a shock to the class of landlords. Pangarker, a leader of peasants in Baroda, Gujarat, addressing the conference, described landlords as social parasites and forcefully advocated denial of farm rent.[11] Desai symbolically burnt an effigy of a landlord and declared that landlordism was turned into ashes. The immediate effect of the conference was that many tenants of Ankola and Gokarn (Kumta) started the campaign of land rent refusal. Landlords, most of whom were also leading Congressmen, ganged up and pressurised the Bombay government to extern Desai. In the last week of May 1940, he was arrested under the Defence of India Rules and externed from Uttara Kannada for a period of five years. [12]

With the externment of Desai, the tenants' movement subsided temporarily; we can even say that the first or the initial phase of the movement ended. In this phase, the Nadors or rich peasants who were educated and who proudly called themselves staunch Congressmen, participated in the activities of the ATRS. Of course, they had no reason to oppose its demand for the application of the Tenancy Act because they were a self-cultivating land-owning class. But the implementation of the Debtors' Relief Act was bound to hit them, for they were also exploiting the poor peasants through usury like the landlords. In fact, the later events showed that they associated with the ATRS in 1940 only to weaken and sabotage the tenancy movement.[13] If they were really interested in the fulfilment of the causes for which the ATRS stood and fought, they would have given a lead to the tenancy movement after the expulsion of Desai, and the poor peasants, obviously infuriated by the cruel treatment meted out to their esteemed leader, would have certainly followed them. There is evidence to show that the Tenancy Act of 1939 which was applied in the districts of Surat, Thana, West Khandesh and in some parts of Dharwad from April 1, 1941, was not applied in Uttara Kannada at the same time because there was no pressure by the tenants. Pointing out the reasons for the piecemeal application of the Act in 1941, Gupte wrote in 1955: "It was then pointed out that tenancy conditions were not uniform throughout the province (now state). In fact, they differed not only from district to district but even from taluka to taluka. It was, therefore, declared by the government that it was inclined to apply the Act in the first instance where the condition of tenants was very bad and the landlords were most extortionate and wanting in public spirit. As a further consideration, it was urged that the Act was most likely to be useful where the number of tenants was in excess of requirements and therefore, the competition to obtain land on rent was keen."[14] Even though the land tenancy system clearly exhibited its sharp exploitative edge in the coastal belt and also in the rice tract of Siddapur taluka of Uttara Kannada, the 1939 Act was not applied in 1941 even in these parts obviously because of the absence of pressure from below. The ATRS became totally inactive after the expulsion of Desai and also due to the half-hearted support of the Nadors. For the same reason, the Debtors' Relief Act of 1939 too was not applied in Uttara Kannada. When the Nadors came forth to support the tenancy movement in 1940, Desai could not suspect their bonafides because he had high regards for them.[15] Unfortunately, the important office-bearers of the ATRS were Nadors only.[16] Thus, the rich peasant class which was to be kept at arm's length, dominated the ATRS which championed the cause of the poor peasants.

In 1940, the ATRS did not criticise the loosely defined provision for personal cultivation and the provision for voluntary surrender of land. Desai who rightly attributed the high level of tenancy as well as the large incidence of fixed-rent tenancy in the coastal region of Uttara Kannada to the preponderance of non-agricultural absentee landlords, thought that these provisions were not likely to cause insecurity and tension for the tenants.[17] But, the events that followed the implementation of the Bombay Tenancy (Amendment) Act of 1946 contradicted his presumption.

In its initial phase, the tenancy movement in Uttara Kannada had a very restricted coverage mainly because it was short-lived. However, in 1940 Desai was not even aware of the problems of tenants in the rice tract of Siddapur,[18] though he played an active role there also in 1947-48.

The ATRS again became active towards the end of 1944 on

account of the initiative taken by S.V. Pikle, Desai's cousin. His close contact with the Halakki Vakkals in Ankola taluka revealed that their heart-rending poverty was the result of their exploitation by the landlords and Nadors.[19] He clearly understood that the latter participated in the activities of the ATRS in 1940 only to sabotage its goal of getting the Debtors' Relief Act applied to Uttara Kannada. Pikle successfully impressed upon the poor peasants who were the members of the ATRS that they had to again pursue a militant course. With Desai's return to Ankola in May 1945, the ATRS started functioning with added zeal. Its members again refused to part with more than one-third of the yield of the paddy land as rent. Massive peasant meetings, processions of peasants singing songs that condemned exploitative landlordism and playing *halage rana vadya* (traditional war-like musical instruments) became regular features of its activities. But now, Desai, who called himself a socialist, did not raise the red flag, lest he should be considered a communist.[20] Now, the effort of the Nadors to get into the ATRS was frustrated so that it became an association of purely poor peasants, which repeated its demand for the execution of the Debtors' Relief Act in Uttara Kannada.[21] After the Tenancy Act of 1939 was applied to the whole of Bombay Presidency on April 11, 1946, by the Congress Ministry which resumed powers on April 1, the ATRS demanded an amendment restricting rent to one-third of the yield of paddy land. However, now also, the ATRS could not visualise the possible adverse consequences of the loopholes of the Act.[22]

The formation of the Kanara District Kisan Sangh (KDKS) on April 22, 1945, marked another important event in the history of the peasant movement in Uttara Kannada.[23] The KDKS led by communists began its activities in the villages of Karwar taluka. Its dynamic leader Dharma Veera Singh was induced by S.V. Parulekar and his companion Godavari, renowned peasant leaders of the Thana district, to organise poor peasants along militant lines. The *chalgeni* tenants and also Mulgenidars cultivating paddy land were its members, the former being numerically important. By January 1946, the KDKS mobilised the poor peasants in many coastal villages of Kumta, Honavar and Bhatkal talukas. Like the ATRS, it also demanded the application of the Agricultural Debtors' Relief Act to Uttara Kannada and instigated its members to refuse to pay more than one-third of the yield of the paddy land as rent. They never opposed each other. But the ATRS led by Desai did not openly associate itself with the KDKS which raised the red flag. There was understanding between them so that the KDKS never stretched its sphere of activity to Ankola taluka. However, the KDKS too failed to visualise the problems quite likely to arise as a result of the loopholes of the Tenancy Act.

Both the KDKS and the ATRS could shake landlordism to its roots in their respective areas of operation. The Congress leaders who could halt the tenancy movement in 1940 just by getting Dinakar Desai externed from Uttara Kannada now found that they could no longer crush it. The representatives of the district in the Bombay Legislative Assembly, though belonging to the landlord class, admitted that some legislative measure for reducing land rent was absolutely necessary to pacify the tenants who were openly resisting high rents. S.S. Shastri (Bhatkal), whose tenants were the members of the KDKS, while supporting the Tenancy (Amendment) Bill of 1946, remarked in the Bombay Legislative Assembly: "The tenants have become conscious of their rights, their poverty and wretchedness. It will be surprising to know that even in the corner of the province, communist agitation is gaining ground and communist doctrines are spreading. If the legitimate aspirations of the tenants are not to be satisfied by some legislation of this kind, then the result would be that communist activities will gather momentum and in short time, we will have to face all the consequences of a serious conflict between the landlords and tenants".[24] R.M. Prabhu (Ankola) said in the Bombay Legislative Assembly that he supported the Tenancy Bill because he was subject to the heavy pressure of the ATRS.[25] Morarji Desai, the erstwhile Minister for Home Affairs and Revenue in the Bombay Presidency too remarked that the relations between the landlords and tenants in Uttara Kannada were strained. He hoped that the 1946 Act might improve these relations.[26]

The Bombay Tenancy (Amendment) Act of 1946 was also intended in the main to provide protection to the tenant like the original Act of 1939.[27] A protected tenant was defined as one who held the land continuously for six years immediately preceding January 1, 1938 or January 1, 1945. A protected tenancy was perpetual so that in the event of the death of a protected tenant, the landlord was to offer to continue the tenancy on the same terms and conditions to the heir of the deceased tenant. It could not be terminated unless the tenant committed offences such as non-payment of rent or sub-division or sub-letting of land. But this apparently complete protection against eviction was severely qualified by the right of resumption given to the landlord by which he could terminate the tenancy of a protected tenant by giving him one year's notice, if the landlord bonafide required the land for personal cultivation. The definition of personal cultivation was as loose as it was in the original Act. Besides protected tenants, the ordinary tenants were also given some security because the lease of an ordinary tenant had to be at least for 10 years' duration. During this period, the ordinary lease could not be terminated except for non-payment of rent, etc., and unlike in the case of protected tenants, it could also not be terminated before the expiry of the lease period on the ground that the land was required by the landlord for personal cultivation. As the Maharashtra Land Reforms Review Committee rightly opined: "Thus, in a sense, the ordinary tenancies were more secure for a period of 10 years than the protected tenancies."[28] Any lease, protected or ordinary, could be terminated by the tenant voluntarily.

The Act maintained that the maximum rent payable by the tenant, either in cash or kind, should not exceed one-fourth of the produce of the irrigated land and one-third in the case of unirrigated land. The landlord had to issue receipts recognising the payment of rent by the tenant.

Before the Tenancy Bill of 1946 became an Act, the landlords in Uttara Kannada made a serious effort to get the district exempted from the operation of the impending legislation.[29] The North Kanara Landlords Association (NKLA), under the leadership of N.G. Shanbhag, gave publicity to the grievances of the class of landlords through the *Kanara Vritta* (Kanara News), a Kannada weekly published from Kumta. The most important argument advanced by the NKLA was that the provision fixing the maximum rent would adversely affect the small landlords. On this ground, M.G. Gopi, a spokesman of the NKLA, argued in the Bombay Legislative Council for exempting Uttara Kannada from the operation of the Act.[30] However, his argument fell on the deaf years of Morarji, who was already impressed by the demand of the ATRS and KDKS. Thus, the opposition of the NKLA which essentially indicated the vulnerable position of the numerically preponderant class of small landlords in the district, carried no weight so that the Tenancy Act was put into effect here.

This is not to say that because of the demand of the ATRS and the KDKS only, the maximum rent was restricted to one-third of the yield in the case of unirrigated land in the entire Bombay Presidency. As stated by Morarji, the working of the 1939 Act in the districts where it was applied in 1941 revealed that the legal provision extending protection to the tenant could not be positively helpful to him without legislative measure fixing maximum rent, and the Amendment Act of 1946 was a step towards removing the shortcomings of the original legislation.[31] He also pointed out that the maximum rates of rent fixed by the Act were based on the popular and generally accepted concept of 'fair' rent. However, the KDKS and the ATRS succeeded in establishing the urgent need for rent reduction in Uttara Kannada so that the efforts of the NKLA to get it exempted from the operation of the Act failed.

In spite of its provision for fixing maximum rent, the 1946 Act turned out to be a landlord-oriented legislation because of its provisions for resumption of land by the landlord for personal cultivation and voluntary surrender. The landlords in Uttara Kannada tried to take advantage of these provisions obviously at the cost of the tenants. Many landlords served notices to their *chalgeni* and *palu* tenants expressing their desire to cultivate the paddy land personally. As a reliable document shows, by December 1947, the number of such notices served in just four coastal talukas exceeded 15 thousand.[32] Even the non-agricultural absentee landlords issued such notices and filed eviction suits against their tenants in the coastal region. Many of them had the ulterior motive of resuming the leased land in the first instance, and let it informally to some other party later. The Tenancy Act of course did not permit such misuse of the provision for personal cultivation, for it stipulated that if the landlord taking possession of the land failed to cultivate it personally within a period of 12 years from the date on which he took possession, he should reinstate the tenant whom he terminated. The landlord could lease out land to any other party only if the concerned tenant expressed his refusal in writing to take back tenancy. However, the landlords owning land in the coastal areas were hopeful of getting parties to cultivate land as informal tenants for high rents who would, in the event of the evicted tenant raising objection or enquiry by the revenue officials entrusted with the responsibility of administering the Tenancy Act, oblige them by saying that they were working on the land as wage labourers under the personal supervision of the landowner and not as tenants. The loosely defined provision for personal cultivation along with the keen competition to lease in land in the coastal villages prompted many non-agricultural non-resident landlords also to file eviction suits.

In the absence of peasant associations, many non-agricultural landlords would have succeeded in misusing the provision for personal cultivation and many a tenant would have failed to take the benefit of the provision that stipulated the maximum rent. The KDKS came out with a timely declaration that all the landless labourers and tenants ready to cooperate with the landlords interested in evicting tenants under the pretext of personal cultivation would have to face social boycott. This declaration was effective in that many landlords could not get tenants to lease in land informally. Owing to the pressure of the KDKS only, the Aval Karkun (a special officer appointed for deciding tenancy disputes) for Kumta taluka, who was inclined to decide eviction suits in favour of landlords, was transferred.[33] Its members argued before the revenue courts that the landlords who filed eviction suits were involved in non-agricultural occupations away from the place of holdings and did not possess even basic farm assets like bullocks and ploughs. These facts were cited by them as clear evidences to the malafides of the landlords. The peasants used to assemble in groups before the revenue courts and help each other by providing witnesses necessary for proving the malafides of the landlords. Because of these reasons, the revenue officials in the coastal talukas who were otherwise inclined to favour the landlord class, became cautious while deciding eviction suits. The NKLA alleged that the revenue officers, under the pressure of the KDKS, rejected the claims of the landlords and decided many eviction suits in favour of the tenants.[34] M.G. Gopi also pointed out in the Bombay Legislative Council that the revenue officials in the coastal talukas of Uttara Kannada were afraid of the Kisan leaders.[35]

Some obstinate landlords in Karwar taluka demanded the pre-determined rent even after the enforcement of the Tenancy Act. The KDKS prevented them from claiming anything more than one-third of the yield. A large gathering of Kisans was held at Karwar on January 3, 1947, in which the office-bearers of the KDKS made a frontal attack on the Tenancy Act for not making provision for sharing of expenses of cultivation by the landlord.[36] Later, in many villages of Karwar taluka, the landlords were compelled to share the farm expensed incurred by the *chalgeni* tenants. The landlords of Karwar were stunned when they received strongly worded notices from the tenants themselves denying the formers' rights over the leased land and also claiming that it belonged to the actual tiller only. Dharma Veera Singh induced tenants to issue such notices. The landlords lodged a complaint with the Bombay government that Singh and other office-bearers of the KDKS were trying to create social disorder. Under their pressure, the Bombay government passed an order on February 27, 1947, externing Singh out of Bombay Province for one year. Besides, he was also required to undergo a year's rigorous imprisonment. Subsequently, all other office-bearers of the KDKS were arrested and detained in Karwar jail under the Bombay Public Security Measures Act.[37] Later, under the leadership of Honnappa Naik, it continued its struggle. While celebrating the first Independence Day at Karwar, the KDKS urged the Government to enact a law to liquidate landlordism. Honnappa Naik too was arrested on August 5, 1948.[38] Thereafter, the KDKS became totally defunct.

The ATRS also frustrated the attempts of many

non-agricultural landlords in Ankola to evict the protected tenants by instituting the social boycott of the peasants ready to cooperate with the former.[39] When some landlords refused to issue receipts recognising the payment of rent by their tenants, the ATRS instructed them to send it through postal money orders. The crop-cutting experiments conducted by the revenue officials for determining the rent payable in kind by the tenant during 1946-47 resulted in over-estimate of the yield of paddy land. Naturally, the tenants could not derive the full benefit of the provision that fixed the maximum rent. The ATRS conducted crop-cutting experiments on its own and sent a report to the Bombay government in which it highlighted the defects of the experiments undertaken by the revenue officials. The government, in turn, directed the Mamlatdar for Ankola to repeat these experiments along the lines suggested by the ATRS. Dinakar Desai and Pikle came down heavily on the drawbacks of the Tenancy Act in the conference of the Halakki Vakkals, the first of its kind in the history of Uttara Kannada, held at Haldipur on May 6, 1947. Amidst the applause of the peasants, Desai raised the slogan "land-to-the-tiller".[40] Subsequently, in the general body meeting held at Ankola on May 29, 1948, the ATRS resolved and declared that unless and until the Bombay government enacted and implemented a legislation to confer ownership on the tenant, it would not stop its struggle.[41]

An important development in this period was the spread of tenancy movement in the rice tract of Siddapur taluka. With the assistance of the local leaders, Desai and Pikle succeeded in setting the poor peasants against landlords. Within the very short period of a year and half, the tenancy movement in the tract became formidable. This is proved by the fact that the Siddapur Taluka Landlords Association, in its representation to the government, on January 6, 1948, alleged that the local revenue officials were, out of fear, obliging the peasants and denying the benefits of the provision for personal cultivation to the landlords.

Thus, the provision for personal cultivation turned out to be a hotly discussed issue in various circles including the judiciary. The judgement delivered by the Civil Court of Honavar on October 13, 1948, in connection with an eviction suit was, however, highly favourable to the landlord class. The tenants involved in the suit, who were the members of the KDKS, argued that they would be deprived of the only source of livelihood in case their landlord would succeed in evicting them. Pointing out that the landlord had already enough land for personal cultivation, they questioned his bonafides in trying to resume the leased lands involved in the suit. The Civil Judge of Honavar completely rejected the arguments of the tenants and observed that "the tenant has no vested right in the property and he cannot insist upon the landlord as it (resumption) would cause him great inconvenience and loss." Quoting the definition of personal cultivation as provided by the Act, the Civil Judge observed: "A person can get the land cultivated through servants under his personal supervision to any extent. It is not necessary that he should personally labour in all the lands cultivated by him...All that we have to see is whether the landlord wants the land to cultivate it personally within the meaning of the Tenancy Act. If that is so, then we cannot attribute malafide to him simply because he already cultivates some other lands."[42] This judgement which was given wide publicity by the NKLA through the *Kanara Vritta* naturally created anxiety and tension for the tenants. If they were not organised and united, their position indeed would have been miserable.

A Spate of Disputes

The enforcement of the 1946 Act was followed by a spate of

disputes. By the end of 1948, as many as 19,078 tenancy cases were filed in Uttara Kannada under various sections of the Act.[43] Though data regarding the number of eviction suits is not separately available, we can reasonably presume that a considerable number of suits were filed by landlords under the pleas for personal cultivation. Of the total number of cases registered under the Act, only 9,167 or 48 per cent were settled by the end of 1948. The decisions in as many as 75 per cent of the cases settled were favourable to the tenants themselves. Undoubtedly, in the absence of organised peasant movement, a large majority of cases would have been decided in favour of landlords and consequently, many tenants eligible for protection in terms of the Act would have been evicted. However, the movement was highly localised. Only the tenants in the purely coastal villages and those in the rice tract of Siddapur were organised. Elsewhere in the district, the tenants unorganised and inarticulate as they were, could be evicted by the lessors who were either landlords or rich peasants by class. Since in the large part of the Malnad region and the interior part of the below-ghat talukas, the bulk of the lessors were resident landowners, it was relatively easy for them to establish their right of resumption. Because of lack of organisation, some tenants in these parts fell a prey to the provision for voluntary surrender of land.[44]

Even in the coastal villages, as available evidence shows, the peasant associations could not prevent all landlords from misusing the provision for personal cultivation. The study by Dandekar and Khudanpur showed that the right of protected tenancy was recorded on 261 plots in 1947-48 in Karwar and Kumta talukas. But by the very next year, while nearly 4 per cent of these plots was resumed by the landlords who actually undertook self-cultivation, on as much as 10 per cent the protected tenants had been evicted and replaced by informal tenants who had agreed to pay more than one-third of the yield as rent.[45] It goes without saying that in the absence of organised tenants' movement, both the resumption and change of tenants would have taken place on much larger scales in the coastal areas. In the villages not covered by the movement, though the misuse of provision for personal cultivation could not be completely ruled out, its extent was less because in these villages the bulk of the lessors belonged to the castes interested in agriculture; most of them undertook self-cultivation on the resumed lands.

The Tenancy Act that conferred certain rights and privileges on *chalgeni* and *palu* tenants, maintained pointed silence about its applicability to the permanent tenants like the Mulgenidars and quasi-permanent tenants like the Nadagidars and Sulgidars. The main intention of the Act of providing protection was not relevant in their case because they enjoyed enough security even before its implementation. Therefore, they were not recognised as protected tenants. As such, there arose a question as to whether these categories of tenants were eligible to get the benefit of the provision that fixed the maximum rent. Of course, the Mulgenidars cultivating gardens in the up-ghat region did not bother to know whether the Tenancy Act was applicable to them because the cash rents they were required to pay according to the original contracts were less than the rates specified by the Act.[46] Since they were rich peasants by class, they opposed the tenancy movement as well as the Act because some of them had leased out their paddy lands to *chalgeni* tenants. However, the Mulgenidars cultivating paddy land in the coastal areas were keen to take the benefit of the provision that stipulated the maximum rent. As they were active participants in the peasant movement, they resisted the demand of their landlords to pay the pre-determined rent and expressed their willingness to pay only at the rate prescribed by the Tenancy Act. Some non-agricultural landlords of Karwar filed suits against their Mulgeni tenants contending that the essential purpose of the Tenancy Act was to provide security to the tenant and as the defendants (i.e., *mulgeni* tenants) had already enough security, they were not eligible to get the benefit of its provision prescribing maximum rates of rent. The Civil Judge of Karwar in his judgement delivered in 1948 upheld the contention of the landlords[47] and the NKLA gave wide publicity to this judgement also. The Mulgenidars, affected by this judgement, filed appeals in the High Court of Bombay, which, while setting aside the judgment of the trial court in 1950, held that Mulgenidars were also tenants and therefore, they were eligible to get the benefit of the Tenancy Act. Thus, now again, the High Court of Bombay maintained its convention of treating the Mulgenidars as tenants with superior rights and not as landlords.[48] Indeed, the Mulgenidars cultivating paddy land could succeed in getting the benefit of the Tenancy Act even by indulging in protracted legal battles because they were bold, assertive and aware of their rights.

The Mulgenidars who sub-let the land experienced much confusion.[49] Since the Tenancy Act made sub-letting of land an offence, it naturally implied that they had to evict themselves. Since such Mulgenidars belonged to the class of landlords, the NKLA represented their case to the government.[50] In order to protect the interests of sub-tenants and tenants having no recorded evidence, the government directed the revenue officials to enter the names of such tenants in the tenancy registers maintained at the village level and to extend to them the rights of protected or ordinary tenancy after making proper enquiry.[51] The peasant associations played a crucial role by inducing their members without any recorded evidence to prove their tenancy status to take the benefit of this arrangement. The rise in the consciousness of bulk of the tenants which was the result of the tenancy movement only, enabled them to insist on the creation of right land records. This becomes obvious from the fact that by May 31, 1948, the number of protected tenants in the district excluding Sirsi and Karwar talukas was 43,819 while there were only 9,597 ordinary tenants.[52]

The peasant movement produced considerable spread effect. The *nadagi* and *sulgi* tenants cultivating coconut gardens in the coastal villages of Kumta and Honavar talukas, the bulk of whom were poor peasants, also paid rents at the rates fixed by Tenancy Act, disregarding the share of the landlords in investment in the leased land. There were some tenants who were Chalgenidars on the paddy land, while they had *nadagi* or *sulgi* rights on the coconut gardens. Since they had become aware of their rights, they naturally resisted the demand of their landlord owning *nadagi* or *sulgi* lands also to pay the contract rents. There were also pure Nadagidars and Sulgidars, but they were not participants in the peasant movement. However, they were inspired by the move of the tenants cultivating paddy land. They too seized the benefit of the provision of the Act that specified the maximum rates of rent. When Morarji went over to Kumta on April 10, 1947, with the sole purpose of inquiring into the problem of agrarian tension, the NKLA argued that since the landlords owning *nadagi* and *sulgi* lands shared the investment costs, they deserved a much greater share in the produce than what the Tenancy Act prescribed.[53] Morarji, however, refused on the ground that it was not possible for the government to enact a separate legislation for Uttara Kannada.

Implementation of the Debtors' Relief Act

The application of the Agricultural Debtors' Relief Act of 1939 (amended in 1946) in June 1946 for cancelling the debts of agriculturists of Kumta and Ankola talukas was welcomed by the poor peasants. The government notification date June 23, 1946, specified that the debtor holding less than four acres of spice garden or six acres of irrigated rice or coconut land or 18 acres of dry land or any land assessed at Rs. 30, whichever was greater in area should be exempted from the payment of his dues, including arrears of rent.[54] The Act was further amended and applied in the whole of Uttara Kannada on February 1, 1947. All the rent decrees, including the rent suits for the years 1945-46 and 1946-47 pending in the revenue courts were transferred to the Debt Adjustment Boards constituted for implementing the Act.[55] Obviously, this Act hit the landlords and rich peasants who included, of course, Nadors.

The Nadors could no longer tolerate the success of the ATRS. They waited for the right moment to oppose it openly. That moment came in 1948 when Desai, who as an active member of the Praja Socialist Party (PSP), started attacking the Congress Party. The Nadors now put forward a flimsy pretext that they could not bear with Desai's criticisms of the Congress. They found in landlords their class allies, who besides being Congressmen like them, were hit by the Debtors' Relief Act. Therefore, together with the landlords they started opposing the ATRS. The association of rich peasants with the ATRS in its early stage made it weak and eventually led to its sabotage.

The Congress Party tried to have its own organisation of tenants. After the introduction of the Tenancy Act, the Kanara District Congress Committee nominated M.M. Jalisatgi (Honavar) as the convenor of the Kanara District Congress Ryot Sangh (KDCRS). One of the leaflets published by the KDCRS itself indicated that it was formed mainly to oppose and counteract the KDKS.[56] However, it could not attract the members of the KDKS. It convened meetings of the tenants in some interior villages of Honavar and Kumta talukas. Indeed, the KDCRS never became a powerful peasant association because, when Jalisatgi started criticising the loosely defined provision for personal cultivation, he was dubbed a communist by his own partymen. Frustrated with this state of affairs, M.G. Shetty and Dayanand Nadkarni who were acting as secretaries to the KDCRS severed their relations with it and joined the ATRS in 1948. In view of the changed circumstances, i.e., decline of the KDKS, importance of the KDCRS and the opposition of the Nadors, the PSP leaders found it necessary to establish a district level peasant organisation. Thus, Ryot Koota (RK) was formed on September 17, 1948.[57] However, it drew its membership from poor peasants in the coastal villages and those in the rice growing tract of Siddapur taluka only. The former members of the KDKS also joined the RK which, however, never raised the red flag like the former.

Problem of *Gazni* Land

The 1946 Act at once disturbed the informal cooperative farming prevailing on the *gazni* land. On account of the provision specifying the maximum rent, there was no incentive left for the landlords owning *gazni* lands to embark upon large scale improvements and reclamations. While many tenants, poor as they were, could not bear the expense of heavy embankments and sluice gates necessary for preventing the entry of saline water into the paddy fields, the relatively well-to-do tenants were prevented by the landlords from investing in the *gazni* land. As reported by D.K. Kunte, the dog-in-the-manger policy of the landlords also came in the way of keeping the *gazni* land in good condition. Therefore, many pieces of *gazni* or Khar land were submerged by sea water and thus became unfit for cultivation.[58] The Bombay Khar Land Act brought into force on December 31, 1948, however, failed to provide any effective solution to the problem. It, of course, provided for the establishment of the Khar Land Development Board which was empowered to prepare schemes for the construction, maintenance and preservation of outer embankments and also to supervise all works in connection with the better cultivation of the Khar land. But it stipulated that the government must contribute 40 per cent of the cost of the schemes, the landlord 10 per cent and the tenant as much as 50 per cent. Like the Tenancy Act, the Khar Land Act also turned out to be a landlord-oriented legislation. Moreover, the implementation of the Act was very slow because by October 24, 1950, the schemes eligible under the Act covered just 852 acres in Kumta and Ankola talukas; it constituted not even 10 per cent of the total area of *gazni* land in the district.[59] In brief, the 1946 Act created more problems than it solved.

The Bombay Tenancy and Agricultural Lands Act of 1948

The Bombay Tenancy and Agricultural Lands Act of 1948 (hereafter BTAL Act) put into effect from December 28, 1948, while repealing the 1946 Act, retained its provisions relating to maximum rent and rights and obligations of the protected and ordinary tenants.[60] It also permitted the landlord to resume the land from the protected tenant for personal cultivation. Its condition that the landlord owning 50 acres of land or more could not resume the land was not likely to make any major dent in Uttara Kannada where more than 93 per cent of the rentiers owned less than 25 acres each.[61] Besides, this condition was not applicable to the provision for voluntary surrender of land. As Dandekar and Khudanpur commented, there existed a real possibility of the landlord, both small and big, resorting to unfair means to induce the tenant into surrendering his land.[62]

The BTAL Act of 1948 made clear provisions regarding the rights and liabilities of the permanent tenants so that the confusion that arose in the wake of the 1946 Act was removed. Since it provided that notwithstanding any agreement or usage, the rent payable by the tenant should not exceed one-third of the yield in the case of the unirrigated land and one-fourth in the case of irrigated land, the Mulgenidars cultivating paddy land in the coastal region of Uttara Kannada faced no problem, and, in this respect, they came on par with the Chalgenidars. However, as Gupte[63] and Patel[64] opined, the 1948 Act treated the permanent tenant on par with the landlord by giving the former the right to sub-lease the land. The landlord could not resume the land from the permanent tenant for personal cultivation. Hence, as the Mysore Tenancy Agricultural Land Laws Committee, popularly called the Jatti Committee, in its report (1958) pointed out, the permanent tenants emerged as a separate category of tenants in the wake of the 1948 Act.[65]

Since the BTAL Act was also biased in favour of the landlord, the RK condemned it since its commencement. It adopted the same strategies which the ATRS and KDKS followed in the past for resisting the efforts of the landlords to evict the protected tenants. That these strategies became successful to a large extent becomes evident from the fact that, during the period 1948-49 and 1952-53, the number of eviction suits filed by the landlords in Uttara Kannada under the pretext of personal cultivation was 1,036 which constituted only 9.6 per cent of the total suits filed by them, whereas the rent recovery cases constituted as much as 60 per cent.[66] The report of investigation by Dandekar and Khudanpur also shows that during the period following the implementation of the 1948 Act, the magnitudes of both the resumption of land and the change of tenants on the plots on which the right of protected tenancy was recorded were less than in the period that followed the 1946 Act. It shows that just within one year, that is, 1947-48 and 1948-49, out of 261 plots under protected tenancy, nearly 4 per cent was actually resumed by the landlords for personal cultivation, while on nearly 10 per cent there was simply change of tenants. However, during the subsequent five-year period (i.e., between 1948-49 and 1952-53), only another 5 per cent of these plots was resumed for personal cultivation and there was changed of tenants on another 7 per cent only.[67] Thus, there is no disputing the fact that in the absence of an active peasant association like the RK, the eviction of protected tenants would have taken place on a much larger scale in terms of the 1948 Act also.

The RK also received stimulus from external sources. In October 1948, Ram Manohar Lohia addressed many meetings of peasants in Uttara Kannada and gave a call to them to continue their struggle.[68] The peasant leaders in Uttara Kannada attended the national level conference of peasants held at Belgaum on April 19, 1949, in which socialists like Jaya Prakash Narayan, N.G. Gore and Lohia provided guidelines for setting poor peasants against exploitative landlords. Inspired by them, the peasant leaders in the district arranged many conferences of peasants and urged for a law for conferring ownership on the tenant. The tenants, too, disappointed and disillusioned with the Tenancy Acts, soon came to know the significance of tenancy abolition. Addressing a large meeting of peasants at Halga on February 22, 1950, Desai pointed out that the Congress government dominated by landlords would never enact any revolutionary law for abolishing tenancy. He called upon the peasants to vote for the PSP in the first national election.[69] The land question thus started assuming political importance in Uttara Kannada.

The election manifesto released by the PSP in the district at the time of the 1952 election took note of the predominance of the small landlords. While making a strong case for tenancy abolition, it argued that the best solution to the land problem in the district consisted in purchasing land from the small landlord by the government at the market rate and selling it to the tenant at a nominal price. Desai, the PSP candidate for the Lok Sabha in Uttara Kannada, tried to impress upon the large class of small landlords that his party had deep concern for them. He was, however, defeated by the Congress candidate by a margin of 48,602 votes.[70] The PSP won only in the Ankola-Karwar constituency, where B.P. Kadam defeated S.P. Gaonkar, a veteran Congressman, and got elected to the Bombay Legislative Assembly. It was construed as an outstanding victory of the RK itself.[71] In other constituencies, however, the Congress candidates got elected because the PSP had established hold only over the poor peasants in the coastal villages and those in some villages of Siddapur taluka, while elsewhere in the district its hold over the voters was considerably weak. Here the landlords and rich peasants, though being a numerical minority, controlled the votes of poor peasants.

The defeat experienced by Dinakar Desai produced some telling effect on his mind. Disappointed with the election result, he ceased to take much interest in the problem of tenants in the district.[72] Therefore, after 1952, the RK, under the leadership of Pikle, continued its fight for a better tenancy reform legislation. Once again, the stimulus for the tenancy movement in Uttara Kannada came from an external source. The RK got inspired by the *adivasi* (tribal) movement organised by Godavari Parulekar in the Thana district. In response to this movement, the government by a notification dated October 17, 1952, reduced the maximum rent to one-sixth of the produce or its equivalent value, irrespective of whether the land was irrigated or not.[63] Pikle thought that the time was now ripe to demand a land-to-the-tiller legislation. On January 7, 1953, a mammoth procession of peasants from Ankola to Karwar (covering a distance of 33 kms.) was organised.[74] The RK in its representation to the Deputy Commissioner of Uttara Kannada strongly pressed for a legislation for the abolition of tenancy. Later, it convened ryot rallies and meetings in many villages in which the demand for tenancy abolition was reiterated.

The BTAL Act of 1948 was extensively amended by the Government of Bombay in 1955, and the Amendment Act (hereafter, the Bombay Act of 1955 or 1955 Act) was formally brought into effect on August 6, 1956. The Maharashtra Land Reforms Review Committee has expressed the opinion that the Bombay Act of 1955 was basically a land-to-the-tiller legislation.[75] The RK claimed that this Act came about because of its pressure only. But Uttara Kannada being a small and remote portion of the Bombay State and the state being the unit for the legislative process, this district could not have dictated the type of legislation needed here, though it could have influenced it. In fact, the protracted struggle of the *adivasis* in Thana district, much closer to Bombay, got more attention than the peasant movement in Uttara Kannada, and exerted pressure on the Congress Ministry in the Bombay State to secure reforms.[76] We can only say that the RK reinforced the cause for which the *adivasis* were relentlessly fighting.

The Bombay Act of 1955 declared that, with effect from April 1, 1957, to be known as Tillers' Day, every tenant, within certain limitations specified in the Act, would be deemed to have purchased the land in his possession free of all encumbrances. But, in Uttara Kannada, the Act was not likely to bring about any substantial transfer of ownership of land to the tenant, because it exempted the small landlord from this provision and permitted tenancy to continue. The small landlord[77] was defined as one who did not own more than an economic holding. The definition of economic holding was such that more than 80 per cent of the landlords in Uttara Kannada secured the right to lease out land.[78] Furthermore, like the earlier legislations, the Bombay Act of 1955 also made provisions for voluntary surrender and the resumption of the leased land for personal cultivation. A tenant could voluntarily surrender the land to the landlord before Tillers' Day. The Act of course fixed a restriction that the landlord could retain the surrendered land for personal cultivation only if his total land after surrender did not exceed one 'ceiling area'. However, the ceiling area was so defined that more than 85 per cent of the landlords in Uttara Kannada could not be affected because here small landlords were predominant. The leased land could be resumed for personal cultivation only to the extent it resulted in the land under personal cultivation being of one ceiling area.[79] The right of the landlord to resume was subject to an additional condition that he could resume only upto 50 per cent of the leased land; at least half the leased land was to remain with the tenant. Thus, the provision for resumption became far more restrictive than it was before. However, the provision for voluntary surrender was not subject to the additional condition that regulated the right of resumption. Therefore, the Act created scope for the landlord to resort to means for pressurising his tenant to surrender tenancy. However, the rent payable by the tenant was drastically reduced. Now all the rent had to be only in cash fixed at just Rs. 20 per acre. This provision gave a sense of achievement to tenants' struggle in Uttara Kannada.

The 1955 Act had practically little or no time to produce its full effect in Uttara Kannada because in November 1956, the district came under the erstwhile Mysore administration. An ordinance issued by the erstwhile Mysore government on March 11, 1957, and subsequently, the Mysore Act No. 13 of 1957, suspended the operation of the provisions of the Bombay Act of 1955 relating to resumption by landlords, purchase by tenants and also voluntary surrender.[80] But the provision that fixed farm rent remained in force in the Bombay-Karnataka region of which Uttara Kannada was a part until October 2, 1965, the date from which the Mysore Land Reforms Act of 1961 was made operative. Thus, during the period of 1957 to 1965, the tenant in the district could not be evicted by the landlord unless he failed to pay rent; nor could he be pressurised to surrender land. The tenant was entitled to get a far greater share in the produce of the leased land than the he did in the past because rent was to be more than Rs. 20 per acre.

Indeed, this situation approximated to the goal of "land-to-the-tiller"[81].

During this period, viz., 1957 to 1965, the agrarian structure in the coastal region of Uttara Kannada underwent significant changes. The NKLA which fought for safeguarding the interests of landlords in the late 1940s and the early 1950s ceased to function after the Bombay Act of 1955 was enforced. The landlords owning land in the coastal villages of the district saw that the wind was certainly blowing against them. Since they could claim only nominal cash rent, the leased land naturally ceased to be a significant source of income for them. As V.T. Hegde observed in 1965: "The landlord completely ceased to take interest in the land and started to seek employment in towns and cities and took up trade or lucrative vocation... This they say, was a blessing in disguise."[82] Thus, not only did the interest of the landlords in the leased land diminish but also there was an increase in the number of non-cultivating absentee landlords. This change in the agrarian structure later facilitated the implementation of the Karnataka Land Reforms Act of 1974.

To sum up, the tenancy movement initiated a process of change in agrarian structure in Uttara Kannada district. The Chalgenidars and *palu* tenants cultivating paddy lands in the coastal villages and the rice-growing tract of Siddapur taluka of the district revolted against the feudal landlords the vast majority of whom belonged to the so-called superior castes. Both the Bombay Tenancy(Amendment) Act of 1946 and the Bombay Tenancy and Agricultural Lands Act of 1948 were landlord-oriented legislations. Many poor peasants in Uttara Kannada would have been proletarianised due to such legislations which favoured the class of landlords in the absence of the organised peasant movements. The Bombay Tenancy and Agricultural Lands (Amendment) Act of 1955 which also came about as a result of the peasant struggles in Uttara

Kannada was essentially helpful to the poor peasants here because it eventually made the leased land an unattractive property for the landlords. The change in the agrarian structure that was initiated in 1940 by the peasant movement was reinforced after the implementation of the Act of 1955. This change manifested itself in the form of considerable increase in the number of non-agricultural absentee landlords in the district, which as will be revealed in the next chapter, facilitated the transfer of the ownership of the leased land to the tenants in terms of the Karnataka Land Reforms Act of 1974.

Notes

1. K.S. Gupte, *The Bombay Tenancy and Agricultural Lands Act of 1948: A Commentary*, Poona, 1955, p. 1.
2. Government of Mysore, *Report of the Mysore Tenancy Agricultural Land Laws Committee*, Government Press, Bangalore, 1958, p. 23.
3. Personal cultivation was defined to mean cultivation of land on one's own account: (i) by one's own labour, or (ii) by the labour of any member of one's family, or (iii) by servants on wages payable by cash or kind but not in crop-share or by hired labour under one's personal supervision or the personal supervision of any member of one's family.
4. Bombay Legislative Assembly, *Debates: Official Report*, Vol. 9, Part 3, July 17, 1946, Government Central Press, Bombay, 1947, p. 85.
5. V.D. Yennemadi, *Papers relating to the First Revision Settlement of Ankola Taluka of the Kanara Collectorate*, The Government Central Press, Bombay, 1917, p. 20.
6. D.V. Ambekar, "Dinakar Desai: An Appreciation", in Gourish Kaikini (ed.), *Dinakara Darshana*, Dinakar Shastyabdi Samiti, Ankola, 1971 (Kannada), p. 10.

7. *The Kanara News*, March 25, 1940, p. 9.
8. The Bombay Agricultural Debtors' Relief Bill received the assent of the Governor-General on 30th January, 1940, and became law. However, it was not implemented anywhere in Uttara Kannada.
9. K.S. Gupte, *op.cit.*, p. 2.
10. *The Kanara News*, April 18, 1940, p.9.
11. Pangarker's speech delivered in Marathi was translated into Kannada by Desai. See the *Kanara News*, May 13, 1940, p. 8.
12. *The Kanara News*, June 30, 1940, p. 9.
13. The Nadors started opposing the ATRS after the Debtors' Relief Act was applied in the entire district in 1947.
14. K.S. Gupte, *op.cit.*, p. 2.
15. As early as 1930, Dinakar Desai had, in one of his poems, paid eloquent tributes to the Nadors for their active participation in the freedom movement. When they came forth to participate in the activities of the ATRS, he wrongly thought they had become altruistic, even unmindful of their class interests. The Nadors did not reveal their true colour in the initial phase because it did not last for more than three months.
16. To facilitate the routine activities of the ATRS, an executive committee of 60 members with Ram Nayak as the President, Dinakar Desai as the Honorary Secretary and Ramchandra Nayak as the Joint Secretary was formed. Both the President and the Joint Secretary were Nadors. See the *Kanara News*, March 25, 1940, p. 9.
17. Dinakar Desai, *Land Rents in Kanara*, The Servants of India Society, Bombay, 1940, pp.3-6.
18. In 1940, Desai wrote, " The problem of tenants is practically confined to the coastal region.... The problem is of little importance in the forest tract because this region contains mostly spice gardens cultivated by owners themselves with the help of hired labour", *Ibid*, p. 1.
19. Pikle met many *Halakki Vakkals* for collecting information for his Ph.D. dissertation on the subject *"The Socio-Economic conditions of the Halakki Vakkals in North Kanara"*. In the

course of his investigation, he found that even the microscopic minority of the *Halakki Vakkals* who owned land were indebted to the Nadors. They had pledged their lands to the usurers and were on the verge of losing the same to the latter as they could not discharge their debt obligations. Pikle was moved by seeing the deplorable condition of the *Halakki Vakkals* to such an extent that he forgot his academic pursuit and started setting them against landlords. See Vishnu Naik, *Before and After Externment*, Raghavendra Prakashana, Ankola, 1982 (Kannada), p. 20.

20. The All India Congress Socialist Forum (1934) of which Dinakar Desai was a member, developed after 1942 a very sharp anti-Communist and anti-Soviet line of propaganda. See R. Palme Dutt, *India Today*, Mani Sanyal, Calcutta 1986, (Sixth Edition), p. 435.
21. The Nadors again expressed their readiness to participate in the activities of the ATRS. But Pikle informed Desai that to entertain them would be a suicidal step. Therefore, they were not allowed to enter the ATRS.
22. The original 1939 Act was in operation in Uttara Kannada from April 11 to November 6, 1946. Though the Congress Ministry applied the Act to the whole Bombay Province as a matter of policy, it had already decided to amend it. Therefore, the revenue officials in Uttara Kannada were not yet directed to identify protected and ordinary tenants. Hence, the landlords here could not evict their tenants. See the *Kanara News*, November 16, 1946, p. 9.
23. Bombay Legislative council, *Debates: Official Report*, Vol. 10, Part 3, September 17, 1946, Government Central Press, Bombay, 1947, p. 100.
24. Bombay Legislative Assembly, *Debates: Official Report*, Vol. 8, Part 5, July 19, 1946, Government Central Press, Bombay, 1946, p. 176.
25. *Ibid*, p. 175.
26. *Ibid*, p. 182.
27. For exposition of the provisions of the Act, see Legal Department, Government of Bombay, *The Bombay Tenancy Act*

of 1939 as modified upto 1st January 1947, Government Central Press, Bombay, 1947, pp. 1-26.

28. Government of Maharashtra, Revenue and Forest Department, *The Report of the Committee Appointed for Evaluation of Land Reforms in Maharashtra*, Government Central Press, 1974, p. 9.
29. The Bill contained a provision that the Act would came into effect in any area on such date as the Government might decide. This provision prompted the NKLA to make an effort to get Uttara Kannada exempted from the operation of the Act.
30. Bombay Legislative Council, *Debates: Official Report*, Vol. 10, Part 9, October 10, 1946, Government Central Press, Bombay, 1947, p. 327.
31. Bombay Legislative Council, *Debates: Official Report*, Vol. 10, Part 8, October 9, 1946, Government Central Press, Bombay, 1947, pp. 226-227.
32. See the leaflet containing the resolutions passed in the meeting of peasants arranged by the Kanara District Congress Ryot Sangh at Madgeri village (Honavar) on January 1, 19948. As explained here, the Congress Party later tried to have its own organisation of peasants after the Tenancy Act of 1946 was put into effect.
33. *The Kanara News*, September 8, 1947, p. 3.
34. See the leaflets containing the resolutions passed in the meetings of landlords held at Kadatoka (Honavar), Salkod (Honavar) and Theppa (Kumta) on December 31, 1947. January 1, 1948, and July 4, 1948, respectively.
35. Bombay Legislative Council, *Debates: Official Report*, Vol. 14, Part 10, April 3, 1948, Government Central Press, Bombay, 1949, p. 315.
36. *The Kanara News*, January 27, 1947, p. 2.
37. Morarji disclosed in the Bombay Legislative Assembly that the office-bearers of the KDKS were arrested under the Bombay Public Securities Measures Act which provided for detention for trial for the maintenance of public safety and tranquillity. See Bombay Legislative Assembly, *Debates: Official Report*, Vol. 12, Part 12, March 4, 1948, Government Central Press, Bombay, 1948, p. 399.

38. Bombay Legislative Council, *Debates: Official Report*, Vol. 15, Part 12, October 21, 1948, Government Central Press, Bombay, 1950, p. 602.
39. Vishnu Naik, *op.cit.*, pp. 37-38.
40. *Ibid.*, p. 25.
41. See the leaflet containing the resolutions of the ATRS passed in its general body meeting held at Ankola on May 29, 1948.
42. In this case, the tenants had succeeded in exposing the malafides of the landlord and defeated his efforts to evict them. The landlord later filed a suit against the decision of the revenue officials. See, for details, the *Kanara News*, November 1, 1948, pp. 1-3.
43. The Government of Bombay, *The Bulletin of the Bureau of Economics and Statistics*, Vol. 3, October 1949, pp. 56-57.
44. Unfortunately, data indicating the total number of voluntary surrenders in the district are not available to us. However, in a single month of April 1947, there were 81 voluntary surrenders in Yellapur, Mundagod and Haliyal talukas. See the *Kanara News*, May 31, 1948, p. 6.
45. V.M. Dandekar and G.J. Khudanpur, *The Working of the Bombay Tenancy Act: Report of Investigations*, Gokhale Institute of Politics and Economics, Poona, 1957, p. 171.
46. *Mulgeni* with cash rent was predominant in the case of garden land. Of course, the cash rents paid by the *Mulgenidars* holding interest-leases which were the outcome of usury were higher than those paid by the genuine *Mulgenidars*. However, the rents paid by all *Mulgenidars* were less than the rates prescribed by the Tenancy Act.
47. *The Bombay Law Reporter*, Vol. 53, 1950, p. 315.
48. In Vaikuntha Raya Vs. Shivaram Bhat (1883), Narayan Vs. Shri Ramachandra Deva (1903), Narayan Vs. Nagappa (1910) and Ganapathi Vs. Nagabhatta (1917), the High Court of Bombay treated *Mulgenidars* as permanent tenants and not as landlords, or even as sublords, *Ibid.*, pp. 315-318.
49. Of course, this confusion was short-lived. The Bombay Tenancy Act of 1946 was repealed by BTAL Act of 1948 which made clear provision regarding the rights and liabilities of

permanent tenants.

50. In its representation dated March 24, 1947, the NKLA argued that the *Mulgenidars* who had sublet should not be evicted. It also suggested that some "fair" share such as one-third or one-fourth of rent be reserved in favour of them.

51. K.S. Gupte, *op.cit.*, p. 46.

52. The task of recording the names of protected and ordinary tenants was completed by May 31, 1948. Data for Sirsi and Karwar talukas were not published. See the *Kanara News*, June 6, 1948, p. 6.

53. *The Kanara News*, June 2, 1947, p. 1.

54. Bombay Legislative Assembly, *Debates: Official Report*, Vol. 15, Part 3, October 4, 1948, Government Central Press, Bombay, 1949, p. 105.

55. *Ibid.*

56. A leaflet dated February 25, 1947, published by the KDCRS.

57. M.G. Shetty, "Fulfillment in spite of Death", in Gourish Kaikini (ed.), op.cit., p. 130.

58. D.K. Kunte was deputed by the Government of Bombay to study the problem of *gazni* land in Uttara Kannada. See Bombay Legislative Assembly, *Debates: Official Report*, Vol. 13, Part 18, October 1, 1948, Government Central Press, Bombay, 1949, p. 859.

59. Khar Land Improvement Schemes were undertaken only in three villages, namely, Bobruwada (Ankola), Bilehongi (Ankola) and Vannahalli (Kumta). See Bombay Legislative Assembly, *Debates: Official Report*, Vol. 17, Part 22, October 24, 1950, Government Central Press, Bombay, 1951, p. 1529.

60. For detailed discussion of the various provisions of the Act, See K.S. Gupte, *op.cit.*, pp. 32-138.

61. See Table 2.5.

62. V.M. Dandekar and G.J. Khudanpur, *op.cit.*, p. 5.

63. K.S. Gupte, *op.cit.*, p. 128.

64. G.D. Patel, *Agrarian Reforms in Bombay*, G.D. Patel, Bombay, 1950, p. 16.

65. The Committee was headed by B.D. Jatti. See the Report of the Mysore Tenancy Agricultural Land Laws Committee, *op.cit.*, p. 24.
66. Many landlords filed suits on flimsy grounds to trouble the tenants. Data regarding the total number of suits settled during the period are not available. However, by 1952-53, over 45 per cent of the suits filed by the landlords were decided against them. See Bombay Legislative Assembly, *Debates: Official Report*, Vol. 25, Part 2, December 14, 1953, Government Central Press, Bombay, 1954, p. 92.
67. Dandekar and Khudanpur, *op.cit.*, p. 171.
68. Vishnu Naik, *op.cit.*, p. 28.
69. M.G. Shetty, *op.cit.*, pp. 131-132.
70. *The Kanara News*, January 22, 1952, p. 1.
71. Vishnu Naik, *op.cit.*, p. 36.
72. Dinakar Desai was on active member of the Bombay Municipal Corporation from 1949 to 1962. Gourish Kaikini, an admirer of Desai himself, wrote in 1958 that he could not devote his time for organising peasants in Uttara Kannada because of his involvement in election politics in Bombay. See Gourish Kaikini, "New Era of Organisation of Peasants", *Janasevaka*, October 8, 1958, pp. 3-8.
73. Vishnu Naik, *op.cit.*, p. 37.
74. *Ibid.*
75. The Report of the Committee Appointed for Evaluation of Land Reforms in Maharashtra, *op.cit.*, p. 12.
76. Godavari Parulekar, *Adivasis Revolt*, National Book Agency, Calcutta, 1975, p. 162.
77. The small landlord was defined as one who did not hold more than an economic holding and whose annual income from all sources including the rent of the land did not exceed Rs. 1,500. An economic holding was equal to 16 acres of dry land or eight acres of seasonally irrigated land, or four acres of perennially irrigated land. See the Report of the Committee Appointed for Evaluation of Land Reforms in Maharashtra, *op.cit.*, p. 16.
78. See Table 2.5.

79. Ceiling area was defined as 48 acres of dry land or equivalent area.
80. On the eve of Reorganisation of States different tenancy laws were in operation in different parts of the erstwhile Mysore State. The Government of Mysore issued an ordinance on March 11, 1957, and subsequently passed the Act No. 13 of 1957 with the intention to enact a common tenancy law. See the Report of the Mysore Tenancy Agricultural Land Laws Committee, *op.cit.*, p. 27.
81. For a variety of reasons, the Mysore Land Reforms Act which was to be passed in 1958 itself was delayed.
82. V.T. Hegde, Inaugural Address delivered on the occasion of the symposium on land problems in North Kanara held on January 29, 1965 at the erstwhile Kanara College, Kumta. See N.R. Rao (ed.), *Land Utilisation and Agriculture in North Kanara*, Planning Forum, Kanara College, Kumta, 1965, p. 13.

4

Tenancy Movement and Legislations in the Post-Reorganisation Period

In this chapter, the impact of the tenancy movements and legislations that operated in Uttara Kannada district during the post-reorganisation period is explained. Before analysing this impact it is necessary here to point out that population in Uttara Kannada started growing since 1951 at a higher rate than in the country as a whole; thus, there was a reversal of the earlier behaviour of growth of population.[1] The necessary data are presented in Table 4.1. It can be seen from Table 4.2 that there was a sharp rise in population in all the up-ghat talukas of the district. No doubt with the establishment of more industries within Uttara Kannada, there was immigration, particularly between 1951 and 1961. However, the increase in population in the Malnad region of the district was also due to the development of infrastructure and the eradication of malarial conditions. Consequently, there was increase in the below-ghat labour supply which was one of the main factors facilitating the development of agriculture in the region.[2]

Table 4.1

Population Changes, 1951-1981

Year	*Population in the District (in lakhs)*	*Rate of Growth (%) in the Preceding Decade*	
		Uttara Kannada	*India*
1951	5.18	17.4	13.3
1961	6.90	33.2	21.5
1971	8.49	23.1	24.8
1981	10.71	26.2	25.0

Note: Comparable with figures in Table 2.3.

Source: Census Reports for various years.

Table 4.2

Population Changes in Up-ghat Talukas of Uttara Kannada, 1951-1981

(population in '000s)

Year	*Haliyal*		*Mundagod*		*Siddapur*	
	Population	*Change (%)*	*Population*	*Change (%)*	*Population*	*Change (%)*
1951	36.5	28.5	15.6	43.1	41.8	22.9
1961	42.7	17.0	26.8	71.8	61.5	47.1
1971	85.1	99.3	44.5	66.0	73.8	20.0
1981	132.5	55.7	58.7	32.0	84.5	14.5
Year	*Sirsi*		*Supa*		*Yellapur*	
	Population	*Change (%)*	*Population*	*Change (%)*	*Population*	*Change (%)*
1951	57.7	34.8	18.6	22.6	19.1	26.5
1961	88.6	53.5	34.4	84.9	30.9	61.8
1971	114.2	28.9	35.8	4.0	44.5	44.0
1981	135.1	10.3	52.2	46.0	60.0	34.8

Note: Change refers to the change in population in the preceding decade.

Source: Census Reports for various years.

There was a phenomenal rise in the prices of garden produce which provided incentive for investment in horticulture. Authentic data indicating this rise are available for selected years (see Table 4.3). The rising profitability of garden cultivation coupled with increase in the below-ghat labour supply facilitated the development of horticulture in Sirsi, Siddapur and Yellapur talukas.[3] Therefore, Mulgeni in these talukas emerged as an instrument of the development of productive forces in agriculture. To put it differently, with the conditions becoming favourable for garden cultivation, the drawbacks associated with the Mulgeni system which at times hindered the development of productive forces in agriculture in the pre-independence period now disappeared and its advantages could be reaped by the tenant-gardeners who were indeed capitalist farmers.

Table 4.3

Price Indices of Important Garden Crops in Uttara Kannada

Crops	*1963*	*1967*	*1971*	*1979*	*1981*	*1983*
Arecanut	100.0	155.0	161.7	172.5	367.3	263.9
Cardamom	100.0	376.4	269.5	1,325.0	785.2	2,236.8
Pepper	100.0	142.1	220.1	585.8	530.6	522.5
Coconut	100.0	185.4	273.8	444.9	639.4	821.4

Source: Derived from the Gazetteer for Uttara Kannada, The Government of Karnataka Publication, Bangalore, 1985, p. 564.

Beginning of Decline of Tenancy Movement

During the late 1950s, the peasant movement in Uttara Kannada subsided obviously because there emerged an institutional setting that safeguarded the interests of the tenants. No sooner was the Bombay Tenancy Agricultural Lands (Amendment) Act 1955 (hereafter the Bombay Act of 1955 or

1955 Act) passed than the peasant leaders in Uttara Kannada too developed the feeling that the purpose of the movement was mainly achieved. Therefore, they diverted their attention to reformist activities like the spread of education. The Kanara Welfare Trust (KWT) founded by Desai and other Praja Socialist Party (PSP) leaders in April 1953 started opening high schools and colleges in different places of the district. They failed to see that the Ryot Koota (RK) still had a useful role to play. Indeed, the RK never made any effort to see whether the poor peasants in the non-coastal villages of Uttara Kannada also paid rent at the rate of Rs. 20 per acre. While the provisions of the earlier legislations fixing the maximum rates of rent had no relevance at all to the large part of the non-coastal region, the provision of the 1955 Act pertaining to rent was indeed quite relevant. The tenants cultivating paddy land in many villages of the Malnad region and also in many interior villages of the below-ghat talukas continued to pay rents in kind, the equivalent cash value of which definitely exceeded Rs. 20 per acre.[4] They could not take the full benefit of the 1955 Act because they were not assertive and bold. Indeed, by organising them at least after the enforcement of the 1955 Act, the RK would have emerged as a powerful district level peasant organisation. The peasant leaders by diverting their full attention to education failed to keep the RK active. They, in fact, started alienating themselves from the peasants.

Since the PSP could not retain its hold over the poor peasants, its candidates were defeated in all the constituencies of Uttara Kannada in the second national election held in February 1957. Interestingly enough, the Congress candidates got the votes of many poor peasants in the coastal region also by claiming the 1955 Act as an instance of the pro-tenant policy of their party and got elected everywhere in the district. Even B.P. Kadam (PSP) who had got elected in 1952, was defeated in 1957 in the Karwar constituency.[5] Knowing rightly that his prospects of winning the election were very bleak, Dinakar Desai did not contest at all, though he canvassed for the PSP.

The PSP leaders failed the peasants even when a right moment came. The need for reviving the peasant movement in Uttara Kannada arose particularly after the report of the Jatti Committee was published in 1958. This committee was appointed by the erstwhile Mysore government to examine the existing laws and practices in regard to the land tenancy system in several areas in the state and also to make recommendations for enacting a common tenancy law. Disregarding the memorandum of the RK which brought out the benefits of the 1955 Act and demanded tenancy abolition, the Jatti Committee, besides making a case for enabling the small landlord to lease out land, recommended that the rent payable by the tenant anywhere in the state, either in terms of kind or cash, should not exceed one-fourth of the gross produce in the case of irrigated land and one-fifth in the case of other lands. In addition, it recommended that the landlord should be enabled to resume the leased land for personal cultivation and the tenant be permitted to surrender land voluntarily.[6] The Mysore Land Reforms Act of 1961 mainly based on these recommendations amounted to undoing the gains of peasant struggle in Uttara Kannada. Being fully involved in reformist activities, the PSP leaders did not organise peasants to protest against them. In fact, during the post-reorganisation period, they attended to the tenancy problem only at the time of elections.

We can even question the utility of the activities of the KWT to the class of poor peasants who also belonged to the backward castes like the Halakki Vakkals and the scheduled castes like the Agers. Of course, the KWT was started with the objective of establishing educational institutions in places where the backward classes lived. It also started publishing since 1954 *Janasevaka*, a Kannada weekly, which emerged as an articulate mouthpiece of the PSP. However, the KWT opened higher educational institutions and not primary schools. The children of the poor peasants used to drop out even before completing their pre-primary education and as such, they could hardly benefit from its work. Furthermore, the KWT did not open any adult education centre to impart even basic education to poor peasants by which they could have access to the information on the modern agricultural practices and articles on radical land reform legislations passed in countries like Japan which were published in *Janasevaka*. Thus, the original object of the KWT of spreading awareness among peasants through *Janasevaka*, was not realised to any extent. In fact, only the rich peasants like the Haviks and Nadors and the landlords amply benefited from the educational work undertaken by the KWT.

Truly speaking, after 1955, the RK could not maintain its independent existence.[7] It became a handmaid of the PSP which frequently used the peasants to display its political strength. For example, on November 12, 1960, the PSP organised a big procession of peasants holding broom-sticks which were found useful by Dinakar Desai to clean the Congress Party.[8] In the 1960s, the PSP leaders arranged many meetings and rallies of peasants and insisted that the State Government must release the forest land for cultivation. More often than not, the tenancy problem which also deserved importance was side-tracked or was only casually referred to.

In 1960, there was a split between Desai and Pikle which was basically ideological. By this time, Pikle had become a Marxist whose public utterances were not liked by the members of the PSP in the district. Since the KWT had embarked on an ambitious plan of constructing educational institutions with the help of donations received from rich people and capitalists in Bombay, Pikle's association with the communist party was considered by Desai and Nadkarni as baneful to its future. Desai successfully pressurised Pikle to resign from the life-membership of the KWT.[9] Thus, by the beginning of 1961, they became two poles apart and Pikle's associations with the RK virtually ended. Ironically, after the split, Pikle too took to the cause of education like Desai, instead of organising the peasants in the district.[10]

The PSP leaders attended to the land problem only as a political strategy at the time of the third national election held in February 1962. They at last saw in the Mysore Land Reforms Bill accepted by the legislature in the first week of November 1961, an attempt to undo what the Bombay Act of 1955 had done in Uttara Kannada. In the meeting held at Ankola on November 17, 1961, they resolved to strengthen the peasant movement.[11] The RK was renamed as North Kanara District Kisan Panchayat (NKDKP). It was also decided to open the offices of the NKDKP in all taluka headquarters and big villages. While canvassing for the PSP candidates, Desai pointed out that the Bill, if allowed to become an Act, would cause rise in land rent in Uttara Kannada where the tenants were required to pay only nominal cash rent in terms of the 1955 legislation. The Congress candidates, under the leadership of Ramakrishna Hegde (Siddapur) and J.H. Shamsuddin (Bhatkal)[12] who were prominent in the state level politics, counteracted by saying that the bill had not yet become an Act and therefore, the fear created by the PSP leaders about it was unfounded. They also promised that after getting elected, they would propose suitable legal provisions for protecting the interests of the tenants in the district. This counter-strategy adopted by the candidates of the ruling party was so effective that they could defeat all the PSP candidates, though the latter gave, this time, a tough fight.[13]

Knowing that his prospects of winning were not bright, Dinakar Desai did not contest the 1962 election also.

After the election, the PSP leaders did not take any interest in organising peasants. The NKDKP remained dormant throughout. Its decision to open offices in all the taluka headquarters and big villages remained on paper only. Therefore, the representatives of the district in Vidhan Sabha safely forgot their promise given at the time of election. Of course, the tenants did not revolt on their own because they were yet to be hit by the Mysore Land Reforms Act of 1961, which though received the assent of the President of India as early as March 5, 1962, was not actually put into effect until October 2, 1965.[14]

Provisions of the 1961 Act

Each of the provisions of the 1961 Act[15] was so defined that it could hardly bring any land-to-the-tiller in Uttara Kannada. Its provision of banning the lease was inapplicable to certain types of landowners like minors, unmarried women, widows, mentally or physically handicapped persons and more importantly, small landlords owning two basic holdings.[16] The definition of basic holding was such that more than 75 per cent of the landowners in the district continued to enjoy the right to create tenancy. The 1961 Act also made provisions for voluntary surrender of land and resumption of the leased land for personal cultivation and provided that the tenant could acquire ownership only on the non-resumable land. However, in the Bombay-Karnataka region, the extent of land that the landlord could retain either by exercising his right of resumption or as a result of surrender was subject to the same restrictions and conditions which were specified in the Bombay Act of 1955. This obviously made the provision for resumption more restrictive because the tenant was to be left with at least half the leased land after resumption. But there existed the possibility of the landlords resorting to unfair means to pressurise or induce their tenants to surrender the land because the extent of land that the landlord could retain as a result of surrender was not subject to this additional condition. Besides, the 1961 Act treated the permanent tenant on par with the landlord. In addition to specifying that the landlord could not resume the land held by the permanent tenant, it permitted the latter to create sub-tenancy.

The 1961 Act also recognised a special category of protected tenants, that is, a tenant of any land held continuously and cultivated personally for a period of not less than 12 years prior to October 2, 1965. However, the position of the tenants who were recognised as protected tenants in the Bombay-Karnataka region in terms of the 1946 Act remained undisturbed. In fact, the 1961 Act did not make any remarkable distinction between the rights and liabilities of the unprotected tenant and those of the protected tenants. Yet, the distinction between the two types of tenants assumed importance in Uttara Kannada later in the context of the Karnataka Land Reforms of 1974. The over-whelming majority of the tenants here could easily secure occupancy rights in terms of the 1974 Act because their names were mentioned in the records of rights as protected tenants since many years.

The 1961 Act did not make any effective provision for reinstating the tenants who were dispossessed in Uttara Kannada in terms of the 1946 Act or in terms of the BT and AL Act of 1948. It, of course, provided that the tenants dispossessed before the 10th day of September 1957, could, within one year of the commencement of the 1961 Act, apply to the land tribunal (constituted under the Act) for the restoration of tenancies. But, there was an overriding provision that the tribunal should not restore tenancies if on the appointed day (i.e., 2 October, 1965) the land was being used for non-agricultural purposes, or if it was leased out to some other tenant, or if it was satisfied that eviction or surrender had taken place in accordance with law. Obviously, to get back their tenancies for those tenants who were dispossessed in the late 1940s or the early '50s remained a distant dream.

The provision of the 1961 Act that fixed maximum rent at once hit the tenants in the district. While the tenants in Uttara Kannada were required to pay only nominal cash rent in terms of the 1955 Act, the 1961 Act clearly stated that notwithstanding any law, custom, usage, agreement or the decree or order of court to the contrary, the maximum rent payable, anywhere in the state by the tenant, either in terms of kind or cash, was to be one-fourth of the gross produce in the case of irrigated land and one-fifth in the case of other lands. Owing to this provision only, the 1961 Act turned out to be a retrogressive measure in the Bombay-Karnataka region when compared to the 1955 Act.

The 1961 Act created discontent among the rank and file of tenants in Uttara Kannada particularly after it took effect from 1965. There was a sudden rise in land rents in the coastal villages of the district. While during the 1962 election the damaging impact of the Act had not yet become a reality, the PSP now got a real chance to capitalise on this discontent and proceeded to set the poor peasants against the Congress Party. The 1967 election was close and Dinakar Desai saw that his political ambition could yet be realised. He and other PSP leaders raised a powerful election campaign both in the coastal and non-coastal regions. In his every speech, Desai pointed out that the 1961 Act was passed by the Congress government only to strengthen the position of the landowning class. The PSP candidates, while seeking the support of the poor peasants in the district, promised that they would strive incessantly for a decidedly better tenancy reform legislation. The strategy yielded rich dividends. Three candidates belonging to the PSP, and two candidates supported by it, were elected to the Karnataka Legislative Assembly.[17] Desai went from the "Ryot Sabha to the Lok Sabha."[18] The Gazetteer for Uttara Kannada (1985) has also attributed the success of the PSP leaders in 1967 election to the support of the peasants.[19] Even Pikle, who had actually led the tenancy movement in the early 1950s, experienced a humiliating defeat in 1967 in the Ankola constituency where Dayanand Nadkarni (PSP) was elected, because, after 1960 the former had completely lost his touch with the peasants.[20]

The resounding success of the PSP in the district was celebrated at Ankola on February 27, 1967.[21] Addressing a large meeting of peasants and attributing the success of his party to their whole-hearted support, Desai promised to inform the Indian Planning Commission about the rise in land rents in Uttara Kannada caused by the 1961 Act.[22] He expressed the hope that the Commission might prevail upon the State Government to make suitable amendment to the Act. Subsequently, on March 5, 1967, the PSP gave a call to the tenants in the district to pay rent at the rate of Rs. 20 per acre only, ignoring the 1961 Act.[23] Dayanand Nadkarni threatened in Vidhan Sabha on July 3, 1967, that should the State Government fail to amend the 1961 Act to reintroduce the rate of rent which the Bombay Act of 1955 specified, there was every possibility of outbreak of 'no-rent campaign' in Uttara Kannada.[24]

The efforts put in by the socialists produced the desired effect. As *Janasevaka* of November 15, 1967, reported, in the cabinet meeting held in Bangalore on November 9, a resolution to amend the 1961 Act along the lines suggested by the PSP leaders in Uttara Kannada was passed. However, no such amendment was effected later on. There was no 'no-rent campaign' in the district either. The peasant leaders, who were deeply involved in party politics, could not think of organising such a campaign at all. Since they did not face the problem of insecurity the peasants also did not revolt on their own. By the time the 1961 Act was put into effect, the agrarian structure in the coastal region was substantially different due to the increased involvement of the bulk of the landlords in the non-agricultural occupations who did not file eviction suits. Since they could resume only 50 per cent of the leased land in terms of the Act, filing eviction suits was a game not worth playing, especially because the holdings leased-in by the tenants were exceedingly small. The landlords, however, filed rent recovery suits against their tenants who refused to pay rent at the rates prescribed by the 1961 Act. By October 1967, more than 30,000 rent recovery suits were filed in the district and nearly 70 per cent of them were filed against tenants in the coastal region.[25] But, the bulk of these suits was not settled immediately due to the procedural delays; they were settled much later in terms of the Karnataka Land Reforms Act of 1974 which fixed rent at a much lower rate. No doubt, the tenants in Uttara Kannada were disappointed with their leaders who failed to keep up their promise given at the time of the 1967 election, but resist they did in payment of rents.

The 1961 Act did not bring about any substantial redistribution of ownership of land in Uttara Kannada.[26] It can be seen from Table 4.4 that during the decade 1961 to 1971, the agrarian structure in Uttara Kannada did not witness any significant change. Of course, the proportion of owned and self-operated holdings to total holdings increased during the decade by just 6 per cent, On the contrary, in Karnataka State as a whole it increased by over 18 per cent. There is every reason to believe that the increase in the proportion of owned holdings to total holdings in Uttara Kannada was mainly due to voluntary surrenders of land. That the proportion of pure tenants to total tenants fell by 22 per cent points during 1961-71 (from 65 to 43 per cent) holds a proof to this fact (see Table 4.4). Pani after making due allowance for definitional discrepancy,

Table 4.4
Changes in Different Kinds of Tenancies in Uttara Kannada and Karnataka

S. No.	*Details*	*Uttara Kannada*		*Karnataka State*	
		1961	*1971*	*1961*	*1971*
1.	Percentage of owned and self-operated holdings to total operated holdings	25.9	31.8	70.2	88.8
2.	Percentage of mixed tenant holdings to total holdings	26.2	21.8	19.2	4.6
3.	Percentage of pure tenant holdings to total holdings	47.9	46.4	10.6	6.6
4.	Percentage of pure tenants to total tenants	64.7	42.6	35.5	58.7
5.	Percentage of marginal and small mixed tenants to total mixed tenants	63.4	61.4	25.1	30.0
6.	Percentage of marginal and small pure tenants to total pure tenants	89.3	88.8	71.5	70.3
7.	Percentage of marginal and small holdings to total holdings	75.8	77.6	43.6	54.1

Note: Marginal holdings = below 2.5 acres; Small holdings = 2.5 - 5 acres.

Source: Derived from Census of India 1961, Vol. 11, Mysore, Part III, and Census of Agricultural Holdings in Karnataka, 1970-71.

states that increase in the proportion of agricultural labourers to total rural workers from 12 per cent in 1961 to 20 per cent in the district was due to proletarianisation of peasants.[27] For obvious reasons, there was no scope for large scale resumption of the leased land by the landlord for personal cultivation.

However, the resident landlords in the areas where the tenants were not assertive and bold succeeded in inducing or pressurising them to surrender the land. This became easy particularly because the peasant movement virtually fizzled out after the 1967 election.

Complete Decline of Tenancy Movement

In the 1971 Lok Sabha election, to the surprise of all opposition parties, Desai experienced a crushing defeat at the hands of the candidate fielded by the Congress Party headed by Indira Gandhi, i.e., the ruling Congress. It indeed became a subject of hot discussion all over Uttara Kannada district. While making a post-mortem of the election result on March 27, 1971, the PSP leaders stated categorically that peasants who stood by Desai in 1967, betrayed him in 1971. They went to the extent of calling them 'ungrateful'.[28]

It needs to be stressed here that in 1952, Desai had the support of the poor peasants residing in the coastal areas and the rice tract of Siddapur. In 1967, he could easily secure the votes of many poor peasants because they were hit by the 1961 Act. Besides, he did not encounter any severe opposition of the rich peasants and landlords because by 1967, he was, in their opinion, an educationist and not *havali manja* as he was in the 1940s. Thus, because of these two reasons, he got elected. But in 1971, the Congress Party (O) headed by S. Nijalingappa, and Jan Sangh, both of which were dominated by landed gentry, sided with Desai. It was nothing short of a political irony that Ramakrishna Hegde who was a party to the 1961 Act, and who was also a bitter critic of Desai in 1967 emerged as his admirer in 1971. Canvassing for Desai, Hegde, in the capacity of the Minister for Finance, Government of Karnataka, declared in Kagal that the State Government would enact a law to abolish tenancy, and *Janasevaka* of March 17, 1971, complimented him. Our discussions with knowledgeable persons in Uttara Kannada revealed that in the 1971 election the support of the landed gentry cost Desai the support of the peasants themselves, though it was he who had pioneered the peasant movement in the district. The peasants voted for the ruling Congress which had already promised to enact revolutionary land legislations.[29] In mid-1971 itself, Desai, discontented with the election result, completely retired from politics. Subsequently, many PSP leaders including Dayanand Nadkarni joined the ruling Congress. In fact, by the beginning of 1972 there was neither tenancy movement nor PSP worth the name in the district.

The phenomenon of the socialists joining the ruling Congress in the early 1970s was not confined to Uttara Kannada only. Many socialists in Dakshina Kannada and Shimoga districts, where also tenancy problem was present, joined the Congress, which by way of taking the wind out of the sails of all political parties with the leftist orientation, proceeded to enact the Karnataka Land Reforms Act of 1974. After joining the Congress, the socialists expedited the enactment and execution of the 1974 legislation. The continuation of tenancy movement in Uttara Kannada during the 1970s as such, became unnecessary. But, in spite of its decline, the movement could be termed as a success because it made a large number of poor peasants conscious of their rights and reap the benefits of the Karnataka Land Reforms Act of 1974.

The Karnataka Land Reforms Act of 1974

The Karnataka Land Reforms (Amendment) Act brought into effect on March 1, 1974, removed the provision of 1961 Act that enabled small landlords, widows and minors, etc., to create tenancy.[30] It permitted only two classes of persons, namely, soldiers and seamen to create tenancy on specific terms and conditions. It outlawed tenancy in the case of all others, for it explicitly laid down that when land would be leased out in contravention to its provisions, and upon inquiry the Tahasildar was satisfied to that effect, he was empowered to issue a declaration forfeiting the land to the State Government.

Later, he could summarily evict any person occupying the land. Thus, after the commencement of the Act only those who cultivated the land personally, became eligible to hold it. The definition of personal cultivation was identical with that provided in the Bombay Act or the 1961 Act. However, as the Act did not permit even the small landlord to resume lands, it could not cause eviction of tenants in Uttara Kannada. Besides, unlike the earlier Acts, the 1974 Act did not make provision for voluntary surrender of land by the tenant to the landlord.

All land (other than the land owned by soldiers and seamen) covered by tenancy arrangements as on March 1, 1974, stood vested in the government. The tenants holding these lands had to file declarations or applications to the land tribunal constituted under the Act. The last date for filing applications before the tribunal was originally December 31, 1974. Later on, it was extended up to June 30, 1979. The frequent extension of the time limit for filing declarations was found necessary to enable the tenants to take the benefit of the law.

The Act was intended to enable the sub-tenants also to claim occupancy rights. Those permanent tenants who cultivated the land personally remained on par with the *chalgeni* tenants in securing occupancy rights. Those who had sub-let land had to lose it to the sub-tenant.

The Act made a provision for helping the tenants who were in arrears of rent in terms of the 1961 Act. The rent recovery suits filed on or after October 2, 1965, and still pending were to be decided in terms of the 1974 Act which fixed rent at ten times the aggregate of the assessment on the leased land. This rate was much lower than that fixed by the 1961 Act. The Act prevented, for a period of fifteen years from the date of acquiring occupancy rights, any transfer of land by sale, gift, exchange or assignment. Any such transfer could be declared null and void by the Tahasildar who was empowered to issue a declaration forfeiting the land to the government. The occupant could, however, surrender the land to the government if he desired to give up cultivation for any reason of his own. The objective of the government was, in this connection, to make it obligatory on the part of the new owners of land to enjoy ownership rights for a minimum period of fifteen years.

For the extinguishment of rights of the landlord in the land vesting in the government, the Act provided for compensation to him. The owners of dry land could receive 200 times the land revenue and the owners of irrigated land could receive 150 times land revenue plus 150 times the truncated water rate applicable to the class of land involved. The manner in which the compensation was to be disbursed was different for different categories of landlords. While the landlords constituting vulnerable groups, viz., minors, widows, unmarried women, mentally and physically handicapped persons and also small landlords having total annual income (including income from land) of less than Rs. 4,800 were to get compensation in lumpsum, other landlords were to be paid partly in cash and partly in bonds covering payment of amount in annual installments over a period of twenty years with interest at 5.5 per cent on the amount outstanding.

Originally, the 1974 Act did not make any provision for helping agricultural labourers. But in a special section of the Act dated January 1, 1979, an agricultural labourer residing in a dwelling house on a land not belonging to him became eligible to apply to the tribunal before December 30, 1979, for the ownership of the dwelling house and the appurtenant site around it which was limited to one-twentieth of an acre. While the benefit of this provision cannot be questioned, one wonders why it was not made much earlier. Since this provision was made as late as January 1979 and the time limit for filing declarations was too short, it was not surprising if many agricultural labourers failed to take its benefit.

Thus, while the 1961 Act was characterised by a tendency to legalise without introducing any change in tenancy relations, the 1974 Act was intended to build up a strong case for enabling tenants to acquire ownership rights.[31] However, it did not make any effective provision for rehabilitating those tenants who were already dispossessed from land before March 24, 1974, either by surrender or by eviction. It was, at least to some extent, like locking the stable doors after the horse had already bolted. Had it not been for the tenancy movement in the Uttara Kannada district, the 1974 Act would have been a useless piece of legislation.

Implementation of the 1974 Act

In order to facilitate the task of conferring ownership on tenants, the 1974 Act made provision for establishing land tribunals. There was to be a minimum of one Land Tribunal for each taluka, consisting of four non-official members appointed by the State Government with an Assistant Commissioner as Chairman. The decision of the tribunal was final and no civil court had jurisdiction to question it. All that an aggrieved party could do was to approach the High Court of Karnataka with a writ petition as permitted by the Indian Constitution.[32]

Table 4.5 gives figures of applications which were decided by the tribunals in favour of tenants in relation to the total number in the district. The tribunals completed the disposal of tenancy cases by the middle of 1982; only 90 cases remained undecided as on September 15, 1984. Out of the 98,886 cases decided by the tribunals, well over 80 per cent were decided in favour of the tenant-applicants. Information supplied by the

Table 4.5

Disposal of Tenancy Cases Under the 1974 Act in Uttara Kannada, 1982

(area in acres)

S. No.	Talukas	Applications Received (1)		Applications Decided (2)		Applications Decided in Favour of Tenants (3)		Percentage of (3) to (2)		Applications Rejected		Balance	
		No.	Areas	No.	Areas	No.	Areas	No.	Areas	No.	Areas	No.	Areas
1.	Ankola	10967	19635	10967	19635	9269	17396	84.5	88.6	1968	2239	Nil	Nil
2.	Bhatkal	9891	15869	9891	15869	7762	12981	78.5	81.8	2129	2888	Nil	Nil
3.	Honavar	19063	25003	19063	25003	15926	22269	83.5	89.1	3138	2734	Nil	Nil
4.	Karwar	15473	20300	15452	19943	12511	17258	81.0	86.5	2941	2685	21	357
5.	Kumta	18147	22741	18122	22452	15452	18012	85.3	80.2	2670	4439	25	290
	Coastal Region	73541	103548	73495	102901	60919	87916	83.0	85.4	12576	14985	46	647
		(74.3)	(49.5)	(74.3)	(50.1)	(75.3)	(51.6)			(70.1)	(43.0)	(15.1)	(16.5)
6.	Haliyal	3538	23662	3538	23632	2736	19828	77.3	83.9	802	3804	Nil	Nil
7.	Mundagod	1473	10367	1473	10367	887	5258	60.2	50.7	586	5109	Nil	Nil

Contd...

Contd...

8. Siddapur	8594	26074	8550	22791	7495	20255	87.7	88.9	1055	2536	44	3283
9. Sirsi	8410	32940	8410	32940	6562	26776	78.0	81.3	1848	6164	Nil	Nil
10. Supa	1681	3635	1681	3635	998	2998	59.4	82.5	683	637	Nil	Nil
11. Yellapur	1739	8947	1739	8947	1339	7299	77.0	81.6	400	1648	Nil	Nil
Malnad Region	25435	105595	25391	102312	20017	82414	78.8	80.5	5374	19898	44	3283
	(25.7)	(50.5)	(28.7)	(49.9)	(24.7)	(48.4)			(29.9)	(57.0)	(48.9)	(83.5)
District	98976	209143	98886	205213	80936	170330	81.9	83.0	17950	34883	90	3930
	(100)	(100)	(100)	(100)	(100)	(100)			(100)	(100)	(100)	(100)

Note: Figures in bracket indicate the percentages to district total.

Source: Information supplied by the Deputy Commissioner's Office, Karwar.

Deputy Commissioner's Office, Karwar, reveals that by September 15, 1984, out of 80,936 cases decided in favour of tenants, occupancy certificates had been issued in 80,517 cases leaving a balance of only 365 cases which were covered by the stay orders of the High Court of Karnataka. It can be seen from Table 4.5 that of the 98,886 applications decided by the tribunals in the district, as many as 17,950 were rejected.[33]

Table 4.5 shows regional variations in the number of applications. The five below-ghat talukas of the coastal region accounted for as much as 74 per cent of the total number of cases decided by the tribunals. They also accounted for over 75 per cent of the cases decided in favour of tenants in the district. This varying impact of the 1974 Act was clearly the result of the regional variations in the extent of tenancy in the district.

According to the Agricultural Census for the year 1970-71 the total area under tenancy in Uttara Kannada was 1,58,285 acres (see Table 4.6). This included, of course, only the recorded tenancy and due to the unfavourable 1961 Act, tenancy was under-recorded in 1970-71. However, as Table 4.6 shows, the total area transferred to the tenant applicants in terms of the 1974 Act exceeded the tenanted area as reported in 1970-71. This was because, thanks to tenants organisation and consciousness, many tenants without any recorded evidence also filed declarations and succeeded in securing the benefit of the 1974 Act.[34]

It can be seen from Table 4.6 that in Kumta taluka in the coastal region and Mundagod and Supa talukas in the Malnad region, the area transferred to the tenant-applicants in terms of the 1974 Act was less than the area under tenancy as recorded in 1970-71. While the chances of tenants being dispossessed in the early 1970s in terms of the earlier Act could not be ruled out, some tenants at least, though possessed recorded evidence, failed to file declarations before June 30, 1979.[35]

Table 4.6

Area Transferred to Tenants Under 1974 Act (as on 15 September 1982)

(in acres)

Talukas	*Area under Tenancy as per 1970-71 CAH*	*Area Transferred to Tenants under 1974 Act*	*Difference Between Area Transferred and Area Under Tenancy*
Ankola	15187	17396	2209
Bhatkal	10986	12981	1995
Honavar	18409	22269	3860
Karwar	16274	17258	984
Kumta	20551	18012	-2539
Coastal Region	81407	87916	6509
Haliyal	15461	19828	4367
Mundagod	9662	5258	-4404
Siddapur	18075	20255	2180
Sirsi	23880	26776	2896
Supa	3358	2998	-360
Yellapur	6442	7299	857
Malnad Region	76878	82414	5536
District	158285	170330	12045

Note: CAH = Census of Agricultural Holdings.

Source: Data for 1970-71 in Census of Agricultural Holdings in Karnataka; data for 1982 in Table 4.5.

The goal of the 1974 Act of bringing land-to-the-tiller was, however, realised to a very large extent in Uttara Kannada. The factors accounting for the successful implementation of the Act can now be listed. In the first instance, the tenancy movement in Uttara Kannada, brought about a sharp rise in the consciousness of a large number of poor peasants. The "land-to-the-tiller" slogan was lingering in their ears. They were keen to take the benefit of the 1974 Act.[36] It was easy for a large number of tenant-applicants to secure occupancy rights because their claims were well supported by the records of rights in which their names were mentioned as protected tenants. The creation of right land records was itself due to the rise in the level of consciousness of the bulk of tenants, which in turn was due to the tenancy movement.[37] Secondly, the Act could be executed smoothly because it remained unopposed by a vast majority of landlords. It is interesting to note that the Bombay Tenancy Act of 1946 which was not radical by any standard was opposed by a class of landlords in Uttara Kannada, while the 1974 Act, a relatively revolutionary law, did not face any opposition. Within a short period of four years, viz., during 1974-78, nearly 72 per cent of declarations filed in the district were decided by the tribunals. Hegde found that the majority of the cases were undisputed cases involving stereotyped decisions of the tribunals.[38] In fact, the implementation of the 1974 Act became easy because it affected several absentee landlords. As already stated, the number of absentee landlords in the coastal region increased during the post-reorganisation period.

The disposal of rent recovery cases in terms of the 1974 Act also helped the ex-tenants. An official report shows that by September 15, 1982, as many as 30,846 cases were settled. The five below-ghat talukas accounted for nearly 72 per cent of these cases. Among the up-ghat talukas Siddapur topped the list where 3,258 cases were filed. Since all rent recovery cases were decided in terms of the 1974 Act which fixed rent at a much lower rate, the ex-tenants who had become defaulters under the 1961 Act stood to gain.

Compared to the number of applications filed by the tenants, the applications from homestead dwellers or agricultural labourers were small in number. The official data show that only 3,739 agricultural labourers filed their applications claiming occupancy rights over the dwelling sites. In Haliyal and Supa talukas, the tribunals did not receive any such application. Of the 1,406 applications disposed of by the tribunals by September 15, 1982, as many as 813 or 37 per cent were rejected. In fact, the landless agricultural labourers were never organised in Uttara Kannada. Dinakar Desai and his peasant organisations hardly attended to their problem. Nor could Pikle, despite his Marxist leanings, provide a corrective to this. Thus, the landless labourers and scheduled castes could hardly take the benefit of the 1974 Act. The ceiling law was of little or no relevance in the district characterised by the predominance of small holdings. Official data show that by September 15, 1984, just 152 acres of land was redistributed under the provisions of the 1974 Act relating to ceilings.

To conclude, during the late 1950s, the tenancy movement in Uttara Kannada started declining. By 1971, the process of decline became complete. But, the impact of the movement could be seen at the stage of the implementation of the Karnataka Land Reforms Act of 1974. The movement was instrumental in raising the level of class consciousness of a large number of poor peasants in the district who benefited from the opportunity of acquiring occupancy rights in terms of the Act.

Notes

1. See Table 2.3.
2. V.T. Hegde, Inaugural Address delivered at the Kanara College, Kumta, on the occasion of the symposium on land problems in North Kanara held on January 29, 1965. See N.R. Rao (ed.), *Land Utilisation and Agriculture in North Kanara*, Planning Forum, Kanara College, Kumta, 1965, p. 13.
3. R.M. Karki, "Plantation crops in North Kanara", in N.R. Rao

(ed.), *op.cit.*, pp. 14-18.

4. The author conducted a field survey in Kekkar, an interior village of Honavar taluka. See for details, G.V. Joshi, *Agrarian Structure and Tenancy Reforms: A Case Study of Uttara Kannada*, Ph.D. Thesis submitted to Karnataka University, Dharwad, 1987, pp. 288-289.
5. For party-wise performance in the election, See Suryanath Kamath (ed.), *The Gazetteer for Uttara Kannada*, The Government of Karnataka Publication, Bangalore, 1985, p. 857.
6. Government of Mysore, *Report of the Mysore Tenancy Agricultural Land Laws Committee*, Government Press, Bangalore, 1958, pp. 46-59.
7. The founders of the RK had, while establishing it in 1948, decided to keep it aloof from the influence of all political parties. See M.G. Shetty, "Fulfillment In Spite of Death", in Gowrish Kaikini (ed.), *Dinakar Darshana*, Dinakar Shastyabdi Samiti, Ankola, 1971, pp. 130.
8. *Janasevaka*, November 23, 1960, pp.1 and 7.
9. *Janasevaka*, November 22, 1961, p. 8.
10. Suryanath Kamath (ed.), op.cit.
11. *Janasevaka*, November 22, 1961, p. 8.
12. Both Ramakrishna Hegde and J.H. Shamsuddin belonged to the landlord class.
13. In 1957, the PSP candidate for the Lok Sabha secured only 25.8 per cent of the total votes polled. In 1962, the PSP supported the independent candidate who secured 39.7 per cent of the votes polled, while the Congress candidate for the Lok Sabha secured 46.7 per cent. See Suryanath Kamath (ed.), *op.cit.*, p. 858.
14. As Rajan writes, "This unusual delay was due to legal and constitutional complexity and bureaucratic foot-dragging at the Central Government level. The Presidential assent was given by the Central Government subject to the passing of certain amendments by the State Legislature, and those amendments took time to get passed". See M.A.S. Rajan, *Land Reforms in Karnataka: An Account by a Participant Observer*, Hindustan Publishing Corporation, Delhi, 1986, p. 20.

15. For exposition of the various sections of the Act, see M.R. Achar and T. Venkanna, *Mysore Land Reforms Act of 1961: A Commentary*, Sarala Prakashana, Bangalore, 1968, pp. 101-285.
16. One basic holding was equal to two standard acres. Standard acre meant one acre of the first class of land or an extent equivalent thereto consisting of one or more classes of land specified according to the formula stated in the Act.
17. Ramakrishna Hegde, knowing the gravity of the situation, chose the Haliyal constituency where there was no agrarian tension. "It was in North Kanara that the Congress met with its debacle comparable to the Madras Rout. Five of the six seats were lost and the sole survivor of the holocaust was Finance Minister, Ramakrishna Hegde", see *The Indian Express*, published from Bangalore, dated February 25, 1967, p. 5.
18. This is in the words of Khadri Shyamanna quoted in Vishnu Naik, *Before and After Externment*, Raghavendra Prakashana, Ankola, 1981, p. 8.
19. Suryanath Kamath (ed.), *op.cit.*, p. 187.
20. Dayanand Nadkarni (PSP) secured 14,352 votes and got elected form Ankola. S.V. Pikle, an independent candidate, secured just 1,843 votes. *Ibid.*, p. 859.
21. See the photo on the subsequent page, *Ibid.*
22. *Janasevaka*, March 8, 1967, p. 5.
23. *Janasevaka*, March 15, 1967, p. 7.
24. *Janasevaka*, October 18, 1967, pp. 6-7.
25. The large number of rent recovery suits was due to unwillingness on the part of several tenants in the coastal villages to pay rent at the rates prescribed by the 1961 Act. It did not mean that there was any 'no-rent campaign' as such.
26. The implementation of the 1961 was tardy. The official machinery for implementing the Act was inadequate because too few Land Tribunals were appointed. They had to follow the same procedure as was prescribed for Civil Courts. In 1968, special Tribunals were abolished and the pending applications had to be dealt with by the Civil Courts. Thus, the grant of ownership rights was subject to large delays of indefinite duration. See M.A.S. Rajan, *The Land Reform Laws in*

Karnataka, Government of Karnataka Publication, Bangalore, 1979, p. 10.

27. Narender Pani, *Reform to Pre-empt Change: Land Legislation in Karnataka*, Indian Institute of Management, Bangalore, 1981 (mimeo), p. 51.
28. Dinakar Desai who secured 90,555 votes in 1952, and who was elected in 1967, secured only 62,783 votes in the 1971 election. See *Janasevaka*, March 31, 1971, p. 7.
29. In mid-1969 during the epoch-making Congress session at Bangalore, the then Prime Minister Indira Gandhi was telling her party colleagues "Land reform is no less important (than agriculture). If we do not act urgently, grave political and economic problems will arise", See M.A.S. Rajan, 1986, *op.cit.*, p. 24.
30. For vivid exposition of the important provisions of the 1974 Act, see Rajan, 1986, *op.cit.*, pp. 25-35.
31. Originally, the Act had determined the amount of purchase price to be paid by the tenant in installments. However, an ordinance dated December 6, 1985, promulgated by the State Government and a subsequent amendment provided for the waiver of premium to be paid by former agricultural tenants to the Government if they owned less than ten acres of dry land or land equivalent to it. Therefore, the provision of the Act that fixed the purchase price lost much of its relevance in the present study.
32. The High Court could only remand the matter to the Tribunal for adjudication afresh after quashing its order. It could not substitute its own decision for the decision of the Tribunal. See S.S. Hegde, "A Critical Study of the Working of Land Tribunals in Uttara Kannada District", LL.M. Thesis submitted to the Karnataka University, Dharwad, 1979 (unpublished), p. 56.
33. The frequent extension of time limit for filing declaration and the campaign carried out by the State Government in the early stage to collect applications tended to produce spurious claims, *Ibid.*, p. 127.
34. This interpretation is supported by Hegde's study, *Ibid.*, p.87.

35. *Ibid.*, p. 89.
36. P.S. Kamath, "Kanara Welfare Trust as Welfare Agency", *Kanara Welfare Trust Silver Jubilee Souvenir*, Ankola, 1979, p.92.
37. S.S. Hegde, *op.cit.*, p. 82.
38. By the end of 1978, 93,225 declarations were filed in the district, 67,800 cases were decided of which 84 per cent were decided in favour of tenant-applicants, See S.S. Hegde, *op.cit.*, pp. 91-92.

5

Summary and Conclusions

The thematic structure of our study is based on the issue of making an investigation into the question of how far the goal of equity traditionally associated with the tenancy reform could be realised in a specific regional context, namely, Uttara Kannada district. Our study indicates that the agrarian structure of which tenancy is a part is the outcome of a long historical process. It also shows that to comprehend all the historical, economic and social factors that shape the dynamics of agrarian structure and also to see to what extent the goal of equity is accomplished, we have to rely mainly on the political economy approach.

The historical documents throwing light on the early agrarian structure in Uttara Kannada district clearly indicate that tenancy had been a dominant feature in a region characterised by the preponderance of landowners belonging to the castes not interested in paddy cultivation as it was highly labour intensive and arduous. The predominance of the fixed-rent tenancy in the pre-reform period in the district was also due to the dominance of urban-based non-agricultural landlords. Even those landowners who were residents preferred to lease out their lands on fixed-rent basis because this system of tenancy relieved them from all concern in cultivation. The British policy of encouraging the money-lending business enabled non-agricultural landlords themselves to acquire predominance as usurers also. Many small and poor peasant proprietors fell into their usurious grip. Being unable to discharge their debt obligations, they parted with their property titles to the usurers and subsequently leased in the same lands which they had owned before. The penetration of the merchant's capital into the countryside both in the Malnad and coastal regions of Uttara Kannada subscribed ultimately to the growth of non-resident parasitic landownership. Indeed, by the beginning of the 20th century, tenancy became a pervasive phenomenon in the district. Since the rise of tenancy in the case of paddy land was due to the caste factor and usury, it was basically feudal in character. However, the predominance of *mulgeni* with cash rent in the case of spice gardens shows that tenancy can be a vehicle for developing capitalism in agriculture. Thus, the study of early agrarian relations in Uttara Kannada has proved important and interesting from various academic points of view.

The extent of tenancy was different in different regions of Uttara Kannada. In the Malnad region, characterised by an unhealthy climate and lower density of population and the predominance of agricultural resident landlords, the extent of tenancy was less than in the coastal region. In the latter region, the density of population was more and the non-agricultural non-resident landlords outnumbered resident landlords facilitating a high level of tenancy. The regional variations in the extent of tenancy assumed a lot of relevance in the context of realising the goal of equity through tenancy reforms in Uttara Kannada.

There was indeed a complex pattern of tenancy in Uttara Kannada. *Mulgeni*, *nadagi* and *sulgi* were permanent lease arrangements found mainly in the case of garden crops. On the contrary, *chalgeni* and sharecropping or *palu* were short-term or temporary tenancy arrangements prevailing in the case of paddy. The predominance of *mulgeni* with cash rent in the case of horticultural crops like arecanut, coconut, cardamom and pepper showed that permanent lease contract was suited for the cultivation of commercial crops which required intensive care and heavy long-term investment by the lessee. The dominance of *chalgeni* with fixed rent in kind in the case of paddy showed that temporary lease contract could cater to the requirements of both the parties (the lessor and the lessee), if the cultivation of the leased land did not necessitate any significant long-term investment. However, tenants in the coastal villages and the rice tract of Siddapur demanded security and reduction in rent when tenancy developed a sharp exploitative edge. The goal of equity was relevant in the case of *chalgeni* and *palu* tenants cultivating paddy lands in the above mentioned areas of the district.

Consequent upon the regional variations in the extent of tenancy, there were regional variations in the terms of tenancy which, in turn, indicated the regional variations in the bargaining of the tenant *vis-a-vis* the landlord in the study region. The terms and distribution of benefits in different forms of tenancy in Uttara Kannada could throw an insight into the relative bargaining strength of the two parties. Of course, the bargaining strength of the two parties, before the outbreak of the peasant movement, was different in different regions. However, historical evidence shows that the division of the produce of the leased land between the two parties in general depended on economic factors like the relative shares of the parties in investment and profitability of cultivation, physical factors like the fertility of the soil, and, in some cases, the social factors like caste. The case of interest-*mulgeni* leases showed that when tenancy was the direct outcome of usury, the bargaining strength of the debtor-cum-tenant *vis-a-vis* the creditor-cum-landlord was weak. While the terms secured by the genuine or normal Mulgenidars were equitable, those involved in interest leases obtained inequitable terms. However, almost perfect security of tenure, fixity of rent and hereditary rights over the leased land were the advantages available to all *mulgeni* tenants. Indeed, they were, by and large, privileged tenants, since the Mulgenidars cultivating only garden lands were either rich peasants or landlords by class.

The significance of our study lies in the fact that it has disproved some of the premises held by the scholars who are generally taken as the proponents of the neo-classical theory. Adams and Rask, for instance, argue that fixed-rent tenants, relatively affluent as they are, are not as susceptible to exploitation as tenants under traditional sharecropping system, i.e., sharecropping without cost sharing by the landlords.[1] However, our study, besides indicating that the affluence of the fixed-rent tenant cannot be taken as a *fait accompli*, has brought to light the fact that if fixed-rent and sharecropping systems exist side by side in respect of the same crop in the same region, the gains shared by tenants under both the systems would be more or less the same, though terms apparently differ. In the large part of the Malnad region of Uttara Kannada, where the bargaining strength of the lessee *vis-a-vis* the lessor was strong, both the *chalgeni* and *palu* tenants cultivating the paddy land secured equitable terms and conditions. On the contrary, in the purely coastal villages, the tenants cultivating paddy land, whether as *chalgeni* or *mulgeni* or *palu* tenant, were rack-rented before they were organised because their bargaining strength was weak. Indeed, in these villages the position of the fixed-rent tenant was as vulnerable as that of the sharecropper.

Our study has also contradicted the hypothesis of Bardhan and Srinivasan that a region with a higher wage rate will tend to have a larger incidence of tenancy.[2] In the Malnad region of Uttara Kannada in the pre-independence period, even though farm wages were high, the extent of tenancy was low. In contrast, in the coastal region where farm wages were low, the extent of tenancy was very large. Instead of positive association between wage rate and the extent of tenancy, there was an inverse relation between wage rate and terms and conditions of tenancy. The contractual terms secured by the tenant cultivating the paddy land in the large part of the up-ghat region were equitable because farm wages were high. Before the outbreak of the tenants movement, the tenants in the coastal belt and the rice growing tract of Siddapur taluka were in a disadvantageous position because farm wages here were low. Thus, one of our important findings is that the bargaining strength of the tenant depends, among others, on farm wages. We have also obtained clear evidence to show that half-share system cannot be taken as a universally existing feature. The historical documents indicating the profiles of the early agrarian structure both in the Malnad and coastal regions of Uttara Kannada showed the co-existence of different crop-sharing arrangements in the same region or even in the same village.

Uttara Kannada witnessed a fairly long peasant movement against tenancy and its oppressive character. The intensity of the movement of course varied from time to time. It was instrumental in making a large number of poor peasants increasingly bold, assertive and aware of their rights. The case of Uttara Kannada also shows that unless the tenants are organised on militant lines, the goal of tenancy reform of providing distributive justice cannot be effectively realised.

On more than one occasion, the stimulus for tenancy movement came from external forces. The external stimulus, however strong, could not have ignited the movement but for the glaring contradictions in the land tenancy system and their greater prevalence in the district than elsewhere. The severity of the exploitative character of the tenancy system was one of the objective conditions leading to the origin and development of the movement. Thus, the movement could acquire a mass base at least in the coastal areas of the district where tenancy was more dominant. It exposed effectively the contradictions between the landlord and the tenant, gave expression to the urgent need for tenancy reform and also created the necessary climate for its implementation.

The Bombay Tenancy Act of 1946 and the BTAL Act of 1948, half-hearted and landlord-oriented as they were, intensified class conflicts. The peasant associations in the district started fighting for a land-to-the-tiller legislation and the peasant movement acquired momentum in the years following the implementation of these legislations. The provision of the Bombay Act of 1955 that specified rent, had it been allowed to operate for long, would have extinguished the class of landlords. The Mysore Land Reforms Act of 1961 became a retrogressive measure which again created discontent among the tenants in Uttara Kannada. The Karnataka Land Reforms (Amendment) Act of 1974, which permitted neither resumption of land even by the small landlord nor voluntary surrender of land, could be smoothly carried out in Uttara Kannada. Indeed, after 1957, the *chalgeni* tenants cultivating paddy land in this region were almost on par with the *mulgeni* tenants because the former also enjoyed security like the latter. Therefore, the acquisition of ownership in terms of the 1974 Act did not turn out to be of any significant change for them except that they were not required to pay any rent after March 1974. Half-hearted and pro-landlord tenancy legislations intensified class conflicts and reinforced the tenancy movement in the district. Being a small part of the state, and the state being the unit for legislative process, this district could not have dictated the type of legislation suited to or aspired by the movements though it could have, to some extent, influenced it. After all, the political forces working in the state as a whole were different from those at the district level. These are the limitations of any movement with a restricted local base. However, when a progressive legislation, or a legislation with progressive components, was at last passed, the movement in the district facilitated the implementation of these progressive components.

The class composition of tenancy movement in Uttara Kannada decided whether the benefits of tenancy reform reached the weaker section of the rural society. After the separation of the Nadors from the movement, the poor peasants remained the backbone of resistance. Their bargaining strength improved greatly after they were organised so that they could appropriate the benefits of the tenancy reform. In the absence of the organised movements, many poor peasants in Uttara Kannada would have been proletarianised in terms of the Bombay Tenancy Act of 1946 and the Bombay Tenancy and the Agricultural Lands Act of 1948. Indeed, the case study of Uttara Kannada shows that poor peasants when led by dynamic leaders can take to a rebellious course and achieve the benefits of tenancy reform.

Even in terms of its restricted scope, the movement started to decline in the post-reorganisation period since 1956, due, at least partly, to the alienation of the peasant leaders from the movement and their involvement in election politics. Dinakar Desai who set a model for others by pioneering the movement diverted his attention to educational activities through the Kanara Welfare Trust (KWT) founded in 1953 and remained its chairman until his death in November 1982. Landlords and rich peasants were the major beneficiaries of the work done by the KWT.[3] After 1956, the PSP leaders attended to the tenancy problem only at the time of elections, and not infrequently, even showed concern for small landlords as a vote-catching device. Gourish Kaikini, an admirer of Desai himself, lamented in 1965 that the peasant movement in the district was losing its vitality because of the complete involvement of the PSP leaders in election politics.[4] After Desai became the member of the Lok Sabha and other peasant leaders the members of the Vidhan Sabha in the 1967 election, they had virtually no time to attend to the problems of tenants seriously. The tenants hit by the 1961 Act were unhappy with their leaders because they took no initiative in organising the 'no-rent campaign' which could have certainly become successful. In a public meeting held at Ambarkodla, the very village where Desai was tremendously applauded in 1940, he had to face a criticism that he had forgotten the peasantry, though by way of reply, he pointed out that none in the district could excel him in the past in arranging ryot rallies and meetings.[5] He failed to see that the peasants resented his open association with their class enemies. Thus, the fact of the matter was that he deserted peasants and peasants in turn deserted him in the 1971 election. The process of decline of the movement which started since 1956 reached completion after the 1971 election. In spite of the later lapses, the movement was a solid achievement, and was actually the basis of a successful implementation of the Karnataka Land Reforms Act of 1974. And with it, the movement ended itself, with none left to identify and solve the problems of the peasantry in the post-reform period.[6]

Not surprisingly, the same factors which accounted for the high level of tenancy and the predominance of the fixed-rent tenancy in the district eventually accounted for the successful implementation of tenancy reform. The feudal landlords in the coastal region and the rice-tract of Siddapur taluka exploited their tenants through rack renting as well as through usury as long as their position remained unchallenged. But, when forces started building up against them in the form of peasant movement and reform, the non-agricultural landlords delinked themselves from the leased land and fully concentrated on their non-farm occupations. Of course, they opposed the movement in the 1940s and also early 1950s. However, the movement was so virulent that their opposition did not last long. Moreover, they could not crush the movement because they had no roots in the village society. In fact, the opposition of the landlords against the tenancy movement died out particularly after the enforcement of the Bombay Tenancy and Agricultural Lands (Amendment) Act of 1955 which resulted in a drastic reduction in farm rent. The Karnataka Land Reforms Act of 1974 could be smoothly implemented in Uttara Kannada because it remained unopposed by a vast majority of landlords. Besides, a very large proportion of the landlords in the district were small lessors who did not possess requisite social and political strength to stall any reform nor could they prevent the tenancy movement from spreading. With the peasant movement gathering momentum, the vulnerability of the small landlords in Uttara Kannada also increased. Thus, the predominance of small and non-agricultural urban based landlords was also an objective circumstance which was one of the factors accounting for the success of both tenancy movement and legislation in Uttara Kannada.

On the contrary, in the neighbouring district of Shimoga, the peasant movement in 1950 could be successfully crushed by the big rural based landlords.[7] Thus, the outcome of the movement in Shimoga was dismal when compared to the peasants' movement in Uttara Kannada, though the demands raised by the latter were far more ambitious.[8] Undoubtedly, the success of the movement in the district could be attributed, among others, to the numerical preponderance of absentee landlords, the majority of whom were also petty owners. Thus, the case study of Uttara Kannada indicates that both the tenancy movement and reform have a greater chance of success in a region characterised by the predominance of small and non-agricultural absentee landlords.

We emphasise here that the peasant organisations in Uttara Kannada never thought of organising the landless agricultural labourers. It does not mean, however, that the peasant leaders were completely unaware of their problems. But they failed to realise that issues confronting tenants and landless labourers were different and even mutually contradictory. Desai, while setting himself on the task of organising tenants in 1940, wrote: "Only the fixation of a fair rent will better the present miserable condition of tenants. An improvement in the condition of tenants will also result in bettering condition of landless agricultural labourers, because a rise in the standard of living of tenants means more employment for the field and other labourers and higher wages for them."[9] He did not visualise the possibility of unwillingness on the part of the tenants to pass on a share of their benefits through higher wages and improved working conditions.

The existence of a class of tenants-cum-agricultural labourers was no doubt one of the peculiarities of the early agrarian structure of Uttara Kannada. But after the introduction of the Bombay Tenancy Acts there was a change in the situation. The proportion of agricultural labourers to total workers in the district increased from 8 per cent in 1931 to 12 per cent in 1951. This increase was partly due to the dispossession of considerable number of tenants under the

Bombay Acts, who subsequently entered the rank of landless agricultural labourers. Obviously, this kind of proletarianisation of peasants took place in those of the villages of the district where there was no strong tenancy movement. Pani, after making due allowance for definitional discrepancy, states that increase in the proportion of agricultural labourers to total rural workers from 12 per cent in 1961 to 20 per cent in 1971 in the district was mainly due to proletarianisation of peasants.[10] This possibility escaped the attention of the peasant leaders. There were no attempts whatsoever to raise the bargaining strength of the agricultural labourers. Historical evidence referred to in Chapter 2 suggests that agricultural wages were very low in the coastal belt. But the landless labourers in the district remained unorganised throughout and their specific problems never figured in the discussions of the peasantry. In 'Jungle Satyagraha' (movement for securing forest land for cultivation) also, the poor peasants were participants. The PSP leaders, while leading this *satyagraha*, described the damage caused to agriculture in Uttara Kannada by the 'reckless and cruel' policy of aforestation pursued by the old Bombay government, and strongly defended the encroachments of forest land by the peasants. While making a case for release of forest land they highlighted the disadvantages of small and uneconomic holdings in the district, but referred to the problem of landless agricultural workers in a very casual manner.

The only legislation that conferred some benefit on the agricultural labourers was the Karnataka Land Reforms Act of 1974 which provided for granting occupancy right over the dwelling sites by them. But this provision was made as late as January 1979, so that its effect was very much limited. The landless received little or no benefit from the land ceiling measure for the reason which needs no repetition.

To conclude, the case study of Uttara Kannada shows the

need for a comprehensive study of agrarian structure in a historical perspective for making a realistic evaluation of tenancy movements and reforms. Indeed, the strengths as well as the limitations of the tenancy movement in Uttara Kannada showed that an organised, bold and assertive tenantry is the pre-requisite for the successful implementation of tenancy reforms.

Notes

1. D.W. Adams and N. Rask, "Economics of Cost Sharing Leases in the Less Developed Countries", *American Journal of Agricultural Economics*, Vol. 50, November 1968, pp. 940-941.
2. P.K. Bardhan and T.N. Srinivasan, "Crop-sharing Tenancy in Agriculture: A Theoretical and Empirical Analysis", *American Economic Review*, Vol. 61, March 1971, p. 50.
3. After the Gokhale Centenary College was started in Ankola in 1966, the local landlords became the donors of the KWT. Hari Pai, a non-agricultural landlord who had pressurised the Government of Bombay to expell Desai from Uttara Kannada in 1940, became an active member of the Governing Body of the College. R.M. Prabhu who had also hated Desai in 1940s appreciated him in 1970. As he wrote, "After joining the Servants of India Society, he started agitation for the causes of the peasants..... He later turned an educationist". See R.M. Prabhu, "Dinakar Desai As I Have Known", in Gourish Kaikini (ed.), *Dinakara Darshana*, Dinakara Shastyabdi Samiti, Ankola, 1971 (Kannada), p. 26.
4. Gourish Kaikini, "Let Political History of the District be Published", *Janasevaka*, November 10, 1965, p. 3.
5. *Janasevaka*, November 4, 1970, p. 6.
6. The post-reform problems such as extremely small sized holdings and the menace of floods are highlighted in G.V. Joshi, *Agrarian Structure and Tenancy Reform: A Case Study of Uttara*

Kannada, Ph.D. Thesis submitted to Karnataka University, Dharwad, 1987, Chapter IV.

7. M.V. Nadkarni, *Farmers' Movements in India*, Allied Publishers, New Delhi, 1987, pp. 19 and 20.
8. The agitation which made the movement in Shimoga famous was on a comparatively trival issue, over a measure called *Kolaga* about which the landlords literally employed double standards, one when they sold or loaned foodgrains and another for tenants while paying rent in kind. Tenants of Kagodu village protested against this in 1950. On the contrary, the tenants in Uttara Kannada demanded rent reduction in stages and ultimately abolition of landlordism itself. *Ibid.*, p. 17.
9. Dinakar Desai, *Land Rents in Kanara*, Servants of India Society, Bombay, 1940, p. 7.
10. Narendar Pani, *Reforms to Pre-empt Change: Land Legislation in Karnataka*, Indian Institute of Management, Bangalore, 1981, (mimeo), p. 51.

Bibliography

Books

Achar, M.R., and Venkanna, T., *Mysore Land Reforms Act: A Commentary*, Sarala Prakashana, Bangalore, 1968.

Bergmann, Theodor, *Agrarian Reforms in India*, Agricole Publishing Academy, New Delhi, 1984.

Beteille, Andre, *Studies in Agrarian Social Structure*, Oxford University Press, New Delhi, 1974.

Bharadwaj, Krishna, *Production Conditions in Indian Agriculture: A Study Based on Farm Management Survey*, Cambridge University Press, London, 1974.

Cohen, S.I., *Agrarian Structures and Agrarian Reform*, Martinus Nijhoff Social Division, Leiden, 1978.

Dandekar, V.M., and Khudanpur, G.J., *Working of Bombay Tenancy Act, 1948: Report of Investigation*, Gokhale Institute of Politics and Economics, Poona, 1957.

Desai, Dinakar, *Land Rents in Kanara*, Servants of India Society, Bombay, 1940.

Dhanagare, D.N., *Peasant Movements in India (1920-1950)*, Oxford University Press, New Delhi, 1983.

Dutt, Palme R., *India Today*, Mani Sanyal, Calcutta, 1986, (Sixth Edition).

Ghose, A.K., (ed.), *Agrarian Reform in Contemporary Developing Countries*, Select Book Service Syndicate, New Delhi, 1984.

Grad, A.J., *Land and Peasant in Japan*, International Secretariat, New York, 1952.

Gupte, K.S., *The Bombay Tenancy and Agricultural Lands Act of 1948: A Commentary*, K.S. Gupte, Poona, 1955.

Hegde, S.S., "A Critical Study of the Working of Land Tribunals in Uttara Kannada District", LL. M. Thesis submitted to the Karnataka University, Dharwad, 1979 (unpublished).

Joshi G.V., "A Comparative Study of Land Reforms in Japan, Taiwan and India (*With Specific Reference to Maharashtra*)", M.Phil. Thesis submitted to the University of Pune, 1980 (unpublished).

Joshi, P.C., *Land Reforms in India: Trends and Perspectives*, Allied Publishers, Bombay, 1975.

Kajita, Masaru, *The Land Reform in Japan*, Agricultural Development Series, Tokyo, 1959.

Klein, Sidney, *The Pattern of Land Tenure Reform in East Asia after World War II*, Bookman Monograph, New York, 1958.

Lenin, V.I., *The Development of Capitalism in Russia*, Progress Publishers, Moscow, 1977 (Fifth Edition).

Lenin, V.I., *The Agrarian Question and the "Critics of Marx"*, Progress Publishers, Moscow, 1979 (Third Edition).

Mellor, John W., and Desai, Gunwant M., (eds.), *Agrarian Change and Rural Poverty: Variations on a Theme by Dharm Narain*, Oxford University Press, New Delhi, 1986.

Myrdal, Gunnar, *The Asian Drama: An Inquiry into the Poverty of Nations*, Vol. I, Allen Lane Penguin Press, London, 1968.

Nadkarni, M.V., *Farmers' Movements in India*, Allied Publishers, New Delhi, 1987.

Naik, Vishnu, *Before and After Externment*, Raghavendra Prakashana, Ankola, 1982 (Kannada).

Pani, Narendar, *Reforms to Pre-empt Change: Land Legislation in Karnataka*, Indian Institute of Management, Bangalore, 1981 (mimeo).

Parulekar, Godavari, *Adivasis Revolt*, National Book Agency, Calcutta, 1975.

Patel, G.D., *Agrarian Reforms in Bombay*, G.D. Patel, Bombay, 1950.

Rajan, M.A.S., *The Land Reform Laws in Karnataka*, Government of Karnataka Publication, Bangalore, 1979.

Rajan, M.A.S., *Land Reforms in Karnataka: An Account by a Participant Observer*, Hindustan Publishing Corporation, Delhi, 1986.

Rao, Hanumantha C.H., *Technological Change and Distribution of Gains in Indian Agriculture*, Macmillan Company, New Delhi, 1975.

Rao, N.R., (ed.), *Land Utilisation and Agriculture in North Kanara*, Planning Forum, Kanara College, Kumta, 1965.

Rastyannikov, V.G., *Agrarian Evolution in a Multiform Structure Society: Experience of Independent India*, Routledge and Kegal Paul, London, 1981.

Rudra, Ashok, *Indian Agricultural Economics: Myths and Realities*, Allied Publishers, Bombay, 1982.

Sen, Sunil, *Agrarian Relations in India (1793-1947)*, People's Publishing House, New Delhi, 1979.

Shivanna, K.S., *The Agrarian System of Karnataka (1336-1761)*, The Director, Prasaranga, Mysore University, 1983.

Thorner, Daniel, *Agrarian Prospect in India*, Delhi University Press, 1956.

Thorner, Daniel, *Shaping of Modern India*, Allied Publishers, New Delhi, 1980.

Todaro, Michael, *Economic Development in the Third World*, Longman Group, New York, 1983.

Warriner, Doreen, *Land Reform in Principle and Practice*, Oxford University Press, Bombay, 1969.

Articles

Adams D.W., and Rask, N., "Economics of Cost Sharing Leases in the Less Developed Countries", *American Journal of Agricultural Economics*, Vol. 50, November 1968.

Ambekar, D.V., "Dinakar Desai: An Appreciation", in Gourish Kaikini (ed.), *Dinakara Darshana*, Dinakar Shastyabdi Samiti, Ankola, 1971, (Kannada).

Appu, P.S., "Agrarian Structure and Rural Development", *EPW*, Vol. 9, No. 39, Review of Agriculture, September 1974.

Bardhan, Pranab, "Variations in Extent and Forms of Agricultural Tenancy-II, *EPW*, Vol. 11, No. 37, September 11, 1976.

Bardhan, Pranab, "Variations in Extent and Forms of Agriculture; Tenancy-II, *EPW*, Vol. 11, No. 38, September 18, 1976.

Bardhan, Pranab and Rudra, Ashok, "Inter-linkage of Land, Labour and Credit Relations", *EPW*, Vol. 13, Nos. 32, 33 and 34, Special Number, August 1974.

Bardhan, Pranab, "Agrarian Class Formation", *The Journal of Peasant Studies*, Vol. 10, No. 1, October 1982.

Bardhan P.K., and Srinivasan, T., "Crop-sharing Tenancy in Agriculture: A Theoretical and Empirical Analysis", *American Economic Review*, Vol. 61, March 1971.

Bhaduri, Amit, "Study in Agricultural Backwardness Under Semi-Feudalism", *The Economic Journal*, Vol. 83, March 1973.

Bharadwaj, Krishna and Das, P.K., "Tenurial Conditions and Mode of Exploitation", *EPW*, Vol. 10, Nos. 5, 6 and 7, Annual Number, February 1975.

Chakraborty, Aparijita, "Tenancy and Mode of Production", *EPW*, Vol. 16, No. 13, Review of Agriculture, March 1981.

Chatterjee, Partha, "Agrarian Structure in Pre-partition Bengal", in A.K. Sen, Partha Chatterjee and Sagata Mukherjee (eds.), *Perspectives in Social Sciences-2*, Oxford University Press, Calcutta, 1982.

Dandekar, V.M., "Agricultural Growth with Social Justice in Over-populated Countries", *EPW*, Vol. 5. Nos. 29, 30 and 31, Special Number, July 1970.

Dandekar, V.M., and Rath, Nilakantha, "Poverty in India-II: Policies and Programmes", *EPW*, Vol. 6, No. 2, January 9, 1971.

Gough, Kathleen, "Peasant Resistance and Revolt in South India", *Pacific Affairs*, Vol. 61, No. 4, Winter 1968-69.

Khusro, A.M., "Economic Theory and Indian Agricultural Policy", in A.M. Khusro, (ed.) *Readings in Agricultural Development*, Allied Publishers, Bombay, 1968.

Ladejinsky, Wolf, "Agrarian Reform in India", *Foreign Affairs,* Vol. 62, January, 1964.

Manor, James, "Pragmatic Progressives in Regional Politics: Case of Devraj Urs", *EPW*, Vol. 15, Nos. 5, 6 and &, Annual Number, February 1980.

Nadkarni, M.V., "Tenants from the Dominant Class: A Developing Contradiction in Land Reforms", *EPW*, Vol. 11, No. 51, Review of Agriculture, December 1976.

Parthasarathy, G., and V. Pothana, "Changing Agrarian Structure in Andhra Pradesh", *Journal of Rural Development,* Vol. 2, No. 6, November 1983.

Patnaik, Utsa, "Capitalist Development in Agriculture", *EPW*, Vol. 6, No. 39, Review of Agriculture, September 1971.

Prabhu, R.M., "Dinakar Desai as I have Seen", in Gourish Kaikini, (ed.), *Dinakara Darshana*, Dinakara Shastyabdi Samiti, Ankola, 1971 (Kannada).

Prasad, H. Pradhan, "Reactionary Role of Usurers' Capital in Rural India", *EPW*, Vol. 9, No. 32, 33 and 34, Special Number, August 1974.

Radhakrishnan, P., "Peasant Struggles and Land Reforms in Malabar", *EPW*, Vol. 15, No. 50, December 13, 1980.

Radhakrishnan, P., "Land Reforms and Changes in Land Systems", *EPW*, Vol. 17, No. 39, Review of Agriculture, September 1981.

Rajapurohit, A.R., "Land Reforms and Changing Agrarian Structure in India" A.R. Rajapurohit (ed.), *Land Reforms in India*, Ashish

Publishing House, New Delhi, 1984.

Rajan, M.A.S., "Working of Land Tribunals in India", in A.R. Rajapurohit, (ed.), *Land Reforms in India*, Ashish Publishing House, New Delhi, 1984.

Rao, Hanumantha C.H., "Agricultural Growth and Stagnation", in A.M. Khusro (ed.), *Readings in Agricultural Development*, Allied Publishers, Bombay, 1968.

Rao, Hanumantha C.H., "Uncertainty, Entrepreneurship and Share-Cropping in India", *Journal of Political Economy*, Vol. 79, June 1971.

Reddy, V.N., and Murthy, C.S., "Backward Castes and Tenancy", *EPW*, Vol. 13, No. 26, July 1, 1978.

Rudra, Ashok, "Class Relations in Indian Agriculture-II", *EPW*, Vol. 12, No. 23, June 10, 1978.

Shanin, T., "The Nature and Logic of the Peasant Economy-I", *The Journal of Peasant Studies*, Vol. I, No. 2, January 1974.

Shanin, T.H., "Land Reform and Agricultural Development in Taiwan", in J. Brown and S. Lin (eds.), *Land Reform in Developing Economies*, University of Hartford, 1968.

Suri, K.C. and Raghavalu, C.V. "Agrarian Movements and Land Reforms", in B.N. Yugandhar (ed.), *Land Reforms in India: Andhra Pradesh----People's Pressure and Administrative Innovations*, Sage Publications, New Delhi, 1996, p. 44.

Thorner, Alice, "Semi-Feudalism or Capitalism? Contemporary Debate on Classes and Modes of Production in India", *EPW*, Vol. 17, No. 49, December 4, 1982, No. 50, December 11, 1982, No. 51, December 18, 1982.

Warriner, Doreen, "Land Reform and Economic Development", in Carl Eicher and Lawrence Witt (eds.), *Agriculture in Economic Development*, Vora and Co., Bombay, 1964.

Wolf, Eric, "On Peasant Rebellions" in T. Shanin (ed.), *Peasants and Peasant Societies*, Penguin Book, Harmondsworth, 1969.

Reports

Bombay Legislative Assembly, *Debates: Official Report*, Vol. 9, Part 3,

July 17, 1946, Government Central Press, Bombay, 1946.

Bombay Legislative Assembly, *Debates: Official Report*, Vol. 9, Part 5, July 19, 1946, Government Central Press, Bombay, 1946.

Bombay Legislative Assembly, *Debates: Official Report*, Vol. 12, Part 12, March 4, 1948, Government Central Press, Bombay, 1948.

Bombay Legislative Council, *Debates: Official Report*, Vol. 14, Part 10, April 3, 1948, Government Central Press, Bombay, 1949.

Bombay Legislative Council, *Debates: Official Report*, Vol. 10, Part 8, October 9, 1946, Government Central Press, Bombay, 1947.

Bombay Legislative Council, *Debates: Official Report*, Vol. 10, Part 9, October 10, 1946, Government Central Press, Bombay, 1947.

Bombay Legislative Assembly, *Debates: Official Report*, Vol. 12, Part 2, February 18, 1948, Government Central Press, Bombay, 1948.

Bombay Legislative Council, *Debates: Official Report*, Vol. 15, Part 2, September 17, 1948, Government Central Press, Bombay, 1949.

Bombay Legislative Assembly, *Debates: Official Report*, Vol. 13, Part 18, October 1, 1948, Government Central Press, Bombay, 1949.

Bombay Legislative Assembly, *Debates: Official Report*, Vol. 15, Part 3, October 4, 1948, Government Central Press, Bombay, 1949.

Bombay Legislative Council, *Debates: Official Report*, Vol. 15, Part 12, October 21, 1948, Government Central Press, Bombay, 1950.

Bombay Legislative Assembly, *Debates: Official Report*, Vol. 17, Part 22, October 24, 1948, Government Central Press, Bombay, 1950.

Bombay Legislative Assembly, *Debates: Official Report*, Vol. 25, Part 2, December 14, 1953, Government Central Press, Bombay, 1954.

The Bombay Law Reporter, Vol. 53, 1950.

Government of Mysore, *Report of the Mysore Tenancy Agricultural Land Laws Committee*, Government Press, Bangalore, 1958.

F.A.O. Association, *Agriculture in Japan*, Tokyo, 1958.

Superintendent of Census Operations, Census of India, 1961, Mysore State, *District Census Hand Book*, North Kanara District, Government Press, Bangalore, 1968.

Government of Maharashtra, Revenue and Forest Department, *The*

Report of the Committee Appointed for Evaluation of Land Reforms in Maharashtra, Government Central Press, Bombay, 1974.

Historical Documents

Buchanan, Francis, *A Journey from Madras through the Countries of Mysore, Canara and Malabar*, Vol. II, Higginbotham and Co., 1870 (Second Edition).

Collins, G.F.S., *Introductory Report to the Re-Settlement Reports of the Sirsi, Siddapur and Yellapur Talukas and the Mundagod Mahal of the Kanara Collectorate*, Government Central Press, Bombay, 1926.

Collins, G.F.S., *Revision Settlement Report of Siddapur Taluka of the Kanara Collectorate*, Government Central Press, Bombay, 1926.

Coplestone, W.E., *The Bombay Forests*, Government Central Press, Bombay, 1925.

Desai, M.J., *Papers relating to the Enquiry into the Current Revision Settlements (Garden Lands) of Siddapur, Sirsi, Yellapur Talukas including Mundagod Mahal of the Kanara Collectorate*, Government Central Press, Bombay, 1937.

MacGregor, G.L., *Papers relating to the Second Revision Settlement of the Karwar Taluka of the Kanara Collectorate*, Government Central Press, Bombay, 1913.

Mathrani, K.P., *Papers relating to the Second Revision Settlement of the Mundagod Mahal of the Kanara Collectorate*, Government Central Press, Bombay, 1942.

Mathrani, K.P., *Papers relating to the Second Revision Settlement of the Siddapur Taluka of the Kanara Collectorate*, Government Central Press, Bombay, 1942.

Mathrani, K.P., *Papers relating to the Second Revision Settlement of the Sirsi Taluka of the Kanara Collectorate*, Government Central Press, Bombay, 1942.

Mathrani, K.P., *Papers relating to the Second Revision Settlement of the Yellapur Taluka of the Kanara Collectorate*, Government Central Press, Bombay, 1942.

Mathrani, K.P., *Papers relating to the Third Revision Settlement of the*

Haliyal Taluka of the Kanara Collectorate, Government Central Press, Bombay, 1947.

Maxwell, R.M., *Papers relating to the First Revision Settlement of the Kumta Taluka of the Kanara Collectorate*, Government Central Press, Bombay, 1918.

Maxwell, R.M., *Papers relating to the First Revision Settlement of the Honavar Taluka and Bhatkal Petha of the Kanara Collectorate*, Government Central Press, Bombay, 1921.

Yennemadi, V.D., *Papers relating to the Second Revision Settlement of the Karwar Taluka of the Kanara Collectorate*, Government Central Press, Bombay, 1917.

The Bombay High Court Report, Vol. 12, 1875.

The Gazetteer of the Bombay Presidency, Vol. 15, Part II, Government Central Press, Bombay, 1883.

The Imperial Gazetteer of India, Provincial Series, Bombay Presidency, Vol. 2, Superintendent of Government Printing, Calcutta, 1909, p. 921.

The Land Assessment and Landed Tenures of Canara: Abstracts of Correspondence between the Board of Revenue and the Collectors of Canara, German Mission Press, Mangalore, 1853.

The Kanara District Congress Committee, *A Statement of the Grievances of the Agriculturists and Landowners of Kanara District with Proposals for their Removal*, Kumta, 1938.

Other Materials

The Indian Express, an English Daily, dated February 25, 1967, published from Bangalore.

"Kanada Vritta" *(The Kanara News)*, a Kannada Weekly published from Kumta (relevant issues).

Janasevaka, a Kannada Weekly (1954-1972) published from Ankola, (relevant issues).

Leaflets published by the NKLA and Peasant Associations in Uttara Kannada.

Index

absentee landownership, 30, 64
Adams, D.W. and N. Rask, 26
agrarian capitalism, 24
  characteristics of, 24, 85-86
  development of, 29
agrarian classes, 24
  categories of, 24
agrarian structure, 21
  concept of, 21-22
Alavi, Hamza, 38

Bardhan, Pranab and T.N. Srinivasan, 34
Bardhan, Pranab, 22, 30
Beteille, Andre, 21, 25, 27, 30
Bhaduri, Amit, 28
Bhardwaj, Krishna and P.K. Das, 34
Bhardwaj, Krishna, 29
Bombay Tenancy and Agricultural Land Act, 1948, 117-24
Buchanan, Francis, 62

capitalism
  in plantation sector, 85-88
Chakraborty, Aparajita, 35
Cohen, S.I., 22
Collins, G.F.S., 63, 68, 75, 80, 81

Dandekar, V.M. and Nilkantha Rath, 23
Dandekar, V.M., 35

Debtors' Relief Act, 1939, 114-16
Desai, Dinakar, 66, 82
Desai, M.J., 75
Desai, Morarji, 64

fixed-rate tenancy, 34
  in Uttara Kannada, 65

gazni land
  in Uttara Kannada
  --- cultivation of, 84-85
  --- problem of, 116-17

Hanumantha Rao, C.H., 22, 26, 27, 34
Hardiman, David, 39
Harris, John, 29

Indian agriculture
  capitalism in
  --- development of, 23

Joshi, P.C., 21, 36

Karnataka Land Reforms Act, 1974, 17, 147-56
khatedars (big landlords), 67
Khusro, A.M., 26

land-to-the-tiller
  ideal of, 23
Lenin, V.I., 29

Mathrani, K.P., 76, 80
Maxwell, R.M., 54, 66, 83
middle peasant, 38
Myrdal, Gunnar, 31
Mysore Land Reforms Act, 1961, 140-45

non-agricultural landlordism, 70

Pani, Narender, 29
Parthasarthy, G. and V. Pothana, 36
Patnaik, Utsa, 23
peasant communism
  in South Asia, 37
peasant movements
  in India
  --- role of leadership in, 38-40
  --- Subaltern School of, 38-39
peasant struggles, 37
plantation sector
  in Uttara Kannada, 85-86
  --- development of capitalism in, 86-88
Prasad, Pradhan H., 28

Rajan, M.A.S., 54
Rastyannikov, V.G., 29
Rudra, Ashok, 24, 30

Sabha, Tanka Bahadur, 39
Seth, A.N., 40
sharecropping tenancy, 66
sharecropping, 28

Suri, K.C. and C.V. Raghavulu, 36

tenancy disputes
in Uttara Kannada
— in the pre-reorganisation period, 110-14
tenancy legislations
in India, 41
in Uttara Kannada
— in pre-reorganisation period, 114-24
— in post-reorganisation period, 140-46, 147-50
tenancy movement
in Andhra, 36-37
in India, 36
in Uttara Kannada, 95
— background of, 96-97
— decline of, 135-40, 146-47
— origin and development of, 98-110
tenancy reforms
justification of, 23
working of, 36-42
tenancy relations
role of caste factor in, 33
role of natural factors in, 26-27
tenancy, 22
and agrarian structure
— approaches to study, 25-26: neo-classical, 25-26; political economy, 26
tenancy (*cont.*)
as semi-feudal/feudal phenomenon, 34-35
extent of
— regional variations in, 30-31
forms of, 27-28
in Karnataka, 17-18
— abolition of, 17
in Uttara Kannada
— types of, 75-83
Thorner, Daniel, 21, 24
Todaro, Michael, 21

Uttara Kannada
agrarian structure in, 61-72, 161
and its agricultural economy, 50-54
population changes in
— in pre-reorganisation period, 68-69
— in post-reorganisation period, 133-34
profile of, 46-49
tenancy in
— factors behind, 62-72, 162
— reasons for the growth of, 67
— regional variations in the extent of, 72-75, 162-63
— types of, 75-84, 164-65
tenancy legislations in, 114-24, 166-70

Uttara Kannada (*cont.*)
- — Bombay Tenancy and Agricultural Land Act, 1948, 117-24
- — Bombay Tenancy (Amendment) Act, 1946, 104-5
- — Debtors' Relief Act, 1939, 114-16
- — Karnataka Land Reforms Act, 1974, 147-56
- — Mysore Land Reforms Act, 1961, 140-45

Uttara Kannada (*cont.*)
- tenancy movement in, 95, 165-66
  - — background of, 96-97
  - — decline of, 135-40, 146-47
  - — origin and development of, 98-110

Wolf, Eric, 38

Yennemadi, V.D., 83

Zagoria, Donald S., 37-38